Design Optimization of Wind Energy Conversion Systems with Applications

Edited by Karam Y. Maalawi

Published in London, United Kingdom

IntechOpen

Supporting open minds since 2005

Design Optimization of Wind Energy Conversion Systems with Applications
http://dx.doi.org/10.5772/intechopen.82111
Edited by Karam Y. Maalawi

Contributors
Mai Ayoub, Gerges Edwar Beshay, Karam Youssef Maalawi, Mohamed Kasem, Ridha Cheikh, Hocine Belmili, Carlo Emanuele Dionigi Riboldi, Alessandro Croce, Stefano Cacciola, Luca Sartori, Vitaly Beresnevich, Janis Viba, Martins Irbe, Emilio Gomez-Lazaro, Estefania Artigao, Andrés Honrubia-Escribano, Sergio Martín Martínez, Dumitru Cristian-Dragos, Gligor Adrian, Francis Xavier Ochieng, Craig Matthew Hancock, Gethin Wyn Roberts, Julien Le Kernec, Youjin Kim, Galih Bangga, Antonio Delgado, Ali El Yaakoubi, Lahcen Amhaimar, Adel Asselman

First published in London, United Kingdom, 2020 by IntechOpen
IntechOpen is the global imprint of INTECHOPEN LIMITED, registered in England and Wales, registration number: 11086078, 7th floor, 10 Lower Thames Street, London, EC3R 6AF, United Kingdom
Printed in Croatia

British Library Cataloguing-in-Publication Data
A catalogue record for this book is available from the British Library

Additional hard and PDF copies can be obtained from orders@intechopen.com

Design Optimization of Wind Energy Conversion Systems with Applications
Edited by Karam Y. Maalawi
p. cm.
Print ISBN 978-1-78984-407-8
Online ISBN 978-1-78984-408-5
eBook (PDF) ISBN 978-1-83880-329-2

An electronic version of this book is freely available, thanks to the support of libraries working with Knowledge Unlatched. KU is a collaborative initiative designed to make high quality books Open Access for the public good. More information about the initiative and links to the Open Access version can be found at www.knowledgeunlatched.org

For EU product safety concerns:
IN TECH d.o.o., Prolaz Marije Krucifikse Kozulić 3, 51000 Rijeka, Croatia, info@intechopen.com or visit our website at intechopen.com.

Meet the editor

Karam Y. Maalawi is a professor of Aeronautics and Mechanics at the Mechanical Engineering Department, National Research Centre in Cairo, Egypt. He holds a Bachelor of Science, a Master of Science and a Doctor of Philosophy in Aerospace Engineering from Cairo University. Dr. Maalawi has published extensively in the field of structural optimization and wind turbine design and performance. Likewise, he has utilized the vast range of knowledge at his disposal to contribute to several research projects regarding aerospace engineering technology, wind turbine structures and renewable energy applications. For his outstanding contributions to his profession, the National Research Centre has recognized him with two awards for engineering sciences. He has been presented with the Albert Nelson Marquis Lifetime Achievement Award and endorsed by Marquis Who's Who as a leader in the aerospace engineering industry.

Contents

Preface

Modern and larger horizontal-axis wind turbines with power capacity reaching 15 MW and rotors of more than 235-meter diameter are under continuous development for the merit of minimizing the unit cost of energy production (*$/Kw.hrs*). In this context, the European Wind Energy Association has set a target of installed capacity of 150 GW by 2030, which would provide about 15% of the European Union's demand of electricity. Such valuable advances in this competitive source of clean energy have made vast research contributions in developing wind industry technologies worldwide.

This book presents recent applications and reflects the importance of new ideas and achievements in the expanding field of design optimization of wind energy conversion systems (*WECS*). Part I comprises the first six chapters dealing with optimal production of electricity from wind energy, multi-disciplinary optimization of wind turbines, aerodynamic and structural dynamic optimization and aeroelasticity of the rotating blades. Part II comprises five chapters considering operational monitoring, reliability and optimal control of wind turbine components. The contributing authors combine analysis, design and optimization of *WECS* with description of the implemented mathematical approaches. All chapters were reviewed for their technical contents as stated by the individual authors. As such, the editor assumes no responsibility for the accuracy and completeness of the typing itself.

Chapter 1 discusses the selection of appropriate solutions for planning the electricity production from wind energy. It presents the different algorithms for obtaining models formulated using artificial intelligence techniques such as recurrent neural networks and long short-term memory neural networks. The optimal model for estimating energy production is developed based on energy production history correlated with different parameters describing the weather conditions. Chapter 2 presents a system-level design procedure based on the combination of multi-body numerical models of wind turbines. The overall design aims at minimizing energy cost through the optimization of all the characteristics of the turbine, and the procedure automatically manages all the simulations required to compute relevant loads and displacements. Examples of such studies include the impact of stall-induced vibrations on fatigue, the development of active/passive control laws for large rotors and the complete definition of 10-20 MW reference turbines.

Considering aerodynamic optimization, Chapter 3 deals with optimal airfoil shape of turbine blades using a genetic algorithm (GA). The main objective is to find the best airfoil shape of higher lift coefficient with reduced drag in the boundary layer. The thrust and power curves are generated by the blade element and free vortex codes with 3D and loss correction. The higher power production is given when the wind turbine blades are designed using the optimized airfoil. Final results indicate that soiled condition does not affect the aerodynamic efficiency of the airfoil due to the positive effect of 3D rotation augmentation. Chapter 4 introduces a structural optimization model aimed at improving blade dynamic performance. Three optimization strategies are developed and tested. The first one is based on minimizing the total structural mass under frequency and strength constraints. The second and third strategies are concerned with reducing the overall vibration level by either minimizing a frequency-placement index or maximizing the natural frequencies and placing them at their target values to avoid large amplitudes and resonance occurrence. Exact analytical methods are applied to calculate the natural modes of vibration, and the functional behavior of the coupled frequencies with the lamination parameters is thoroughly investigated and discussed. Other considerations on the process of wind turbine blade design are given in Chapter 5, including the definition of wind turbine loads, selection of a suitable material and creation of an appropriate structural model. A brief discussion on the finite element method and wind turbine aeroelasticity is also given. Chapter 6 considers an approximate method for analyzing the interaction between wind flow and flat blades, allowing synthesis and optimization of wind energy conversion systems without using space-time programming procedures. The method has been tested by computer simulation and experiments in a wind tunnel for determining the applied aerodynamic lift and drag forces. Optimization of system parameters is made to increase the efficiency of energy transformation by changing the orientation of the blade's working surface relative to airflow during rotation of the rotor. Serviceability and main advantages of the proposed method are confirmed by experiments with physical model of an airflow device.

The second part of the book deals with system operational monitoring, reliability and optimal control. The importance of monitoring long composite blades as a key role in power generation is demonstrated in Chapter 7. An overview relating blade testing to Campbell diagrams and non-contact sensors is also addressed as part of blade optimization. Based on design loads under IEC 61400-23 standards, the chapter explores various contact and non-contact sensors for design validation as well as their exploratory use in a three-tier structural health monitoring framework for the blade's operational performance monitoring. The chapter also includes a case study in the non-contact use of Ground-Based Radar in the optimal design of blades and real-time in-field monitoring using condition parameters. Chapter 8 focuses on Operation and Maintenance (O&M) of large wind turbines as a key factor to improve reliability and availability. It has been indicated that Condition-Based Maintenance is a preferred approach for the early detection and diagnosis of critical faults occurring in the induction generator and gearbox; the biggest contributors to failure rates and downtime of wind turbines. Current Signature Analysis is also introduced as a cost-effective and non-intrusive technique that can monitor both mechanical and electrical faults within the induction generator, as well as bearing- and gearbox-related faults.

Chapter 9 concerns the application of reliability theories on wind turbines individually and wind turbine fields as a whole to ensure maximum utilization of available wind power. It develops a reliability/availability-based approach to select appropriate wind turbine types for a wind farm considering site-specific wind speed patterns. The established computer software has the advantage over analytical approaches in calculating the reliability of any complex system, regardless of number of components and where or how they are connected, quickly and accurately with no need for a professional user. The developed methodology has shown to be helpful in the proper selection of wind turbine fields according to the history of wind turbine components and site specifications.

Considering next optimal control strategies, Chapter 10 outlines the different approaches utilized in modeling the major components of a *WECS*. Both aerodynamic and structural modeling of a wind turbine along with a description of the steps used to derive a Linear Time Invariant model are discussed. Thereafter, the chapter introduces models of the electrical actuators in the three phases and Park phases reference frames. The pulse width modulation control strategy, the power converters and the pitch actuator models are also studied. Chapter 11 presents a robust strategy using a nonlinear feedback control technique based on Lyapunov theory. The proposed approach aims to enhance robustness of the wind turbine control scheme, where the permanent magnet synchronous generator has been selected as a case study. It is shown that such a strategy allows for efficient operation of the wind turbine in the standalone operating mode, offers a non-linearity handling of the *WECS* and guarantees a maximum wind power harvesting and robustness against critical working conditions. The chapter also presents numerical simulations within Matlab/SIMULINK environment, which demonstrates the effectiveness and benefits of the proposed methodology.

I wish to express my gratitude and thanks to all authors and other individuals who have contributed to this book and made it possible. I am thankful for the help and support of IntechOpen staff members, particularly Senior Commissioning Editor Ms. Ana Pantar and Publishing Process Managers Mr. Josip Knapic, Ms. Dajana Pemac and Ms. Maja Bozicevic.

Karam Y. Maalawi
Professor of Aeronautics & Mechanics,
National Research Centre,
Cairo, Egypt

Section 1

Optimal Energy Production and Optimization of Wind Turbines

Chapter 1

Optimal Energy Production Planning in Power Systems with Wind Energy Sources

Cristian-Dragos Dumitru and Adrian Gligor

Abstract

Due to its intermittent nature, the optimal production of electricity from wind energy represents a real challenge for nowadays power systems. Whether isolated or grid-connected systems are considered, wind power sources can be profitable, but their intermittent output may lead to problems in terms of power quality and increased costs related to the operation of the grid and to the production of energy. This chapter discusses the choice of the most appropriate solutions for planning the electricity production from wind energy based on different algorithms for obtaining models based on principles used in artificial intelligence techniques such as recurrent neural networks (RNNs) and long short-term memory (LSTM) neural networks. We discuss the situation of obtaining the optimal model for estimating energy production based on a criterion or on multiple criteria: energy production history or energy production history correlated with different parameters describing the weather conditions.

Keywords: energy planning, wind energy, optimal wind energy integration, forecasting, artificial intelligence, RNN, LSTM

1. Introduction

Due to the last years' awareness of climate changes and traditional energy resource depletion, renewable energies started to play a key role in the nowadays electricity market. However, transition to these types of primary resources leads to many challenges to be solved. From the power grid optimal operating point of view, one of the most important issues to be solved is the intermittent and, in some cases, unpredictable availability of the primary energy resources.

According to Ref. [1], in 2017, wind represented the renewable resource with the highest impact on European Union electricity production. The mentioned source reports wind energy with the most important contribution to EU-28 gross electricity consumption with a 30.7% ratio. As reported in Ref. [2], worldwide not including Europe, installed wind power capacities have followed also an ascending trend, being recorded at the beginning of 2017 about 378 GW operational installed power generation capacities, which is rated to about 78% from the total installed capacity. Meanwhile, the wind power is concentrated in few regions. Over 60% is allocated to tree states: China, the USA, and Germany, while most of the countries shares 16%.

The mentioned wind energy development was mainly possible not only due to fiscal facilities developed by many countries as a measure to reduce pollution but also due to some advantages of the wind power such as the lowest priced from the nowadays renewable energy technologies and reduced impact on the land on the installed sites.

The economic advantage on the electricity sale side could be in some situations overcome by the higher initial installation costs compared to the conventional power generation [2]. However, this aspect in future may suffer changes, as the market indicates a decrease in terms of required initial investments. Other features of the wind energy constitute major challenges in its adoption at a higher penetration level. The first one that has to be mentioned is its intermittent availability that is not correlated with the electricity demand and aspect that cannot be overcome simply by installing batteries due to the economic and technical unfeasibility.

The list of attributes can continue with other relevant ones such as the impossibility of this type of energy to be stored, the wind can only be harvested at specific parameters for electricity production, limited proper geographical locations with good wind potential, and in many situations located far away from the consumption centers.

By correlating the wind energy share from the total installed power capacities with the previously mentioned wind characteristics, one can found that power grid can significantly be affected in the case of wind speed changes in terms of network balancing and managing the power flows [3], or in a general speaking term, the stability of the power system can be affected [4]. To overcome this issue, complementary source of energy must be available such as energy storage devices (batteries), pumped-storage hydroelectricity stations, or any traditional electricity generation facilities. In order to operate correctly, with any of the mentioned solutions, due to their technical limitations such as response in time and/or power reserve, forecasting of the wind energy sources availability is a key factor on relying in a sustainable way on this type of energy.

The scope of the next sections is to study different time series modeling methods, recorded in literature, for forecasting production of electrical energy from wind. Two different models are considered: recurrent neural networks (RNNs) and long short-term memory (LSTM) neural networks. The studied time series is the production of electrical energy from wind energy of a national energy system during March 2018.

2. Wind model analysis

In order to understand the behavior of the wind power generation units, a theoretical approach on the mathematical wind model description has to be briefly developed.

One of the most relevant factors in wind modeling description relates to the wind speed.

In order to evaluate the power and electricity generated by a wind power plant, the wind speed analysis is performed by using elements of statistical processing of the measurement data, taking into account its random character. In order to be able to determine the power generated by a wind group, we need an estimation of the wind speed at the height h of the turbine rotor. These data can be obtained by placing an anemometer at some reference height h_{ref} (e.g., 10, 30, or 50 m above the zero wind level). There are two well-known formulas for calculating the wind speed at certain height h [5, 6]:

a. The logarithmic expression:

$$v = v_{ref} \frac{\ln \frac{h}{z_0}}{\ln \frac{h_{ref}}{z_0}} \; [m/s] \tag{1}$$

b. The power law expression:

$$v = v_{ref} \left(\frac{h}{h_{ref}} \right)^{\alpha} \; [m/s] \tag{2}$$

where v is the wind speed at height h above the zero wind level; v_{ref} is the wind speed at h_{ref} (e.g., height of anemometer); and z_0 is the roughness length depending on the surface roughness of a given site (m).

Comparing the above two formulas, the power exponent "α" can be calculated from:

$$\alpha = \frac{1}{2.725 - \ln z_0} \tag{3}$$

The average value of v_m speed, relative to a time period T, is calculated with Eq. (4):

$$v_m = \frac{1}{T} \int_0^T v_i(t) dt \; [m/s] \tag{4}$$

where $v_i(t)$ represents the instantaneous wind speed.

This average wind speed can be hourly, daily, weekly, monthly, quarterly, semi-annual, or annual depending on the reference time interval T, but as a basis of analysis, the hourly average from which the other average values can be determined is used. Despite the observed hourly, daily, and annual average values, the measured wind speeds can vary significantly at different times, at different places, and at different heights relative to the ground.

Therefore, it is very difficult to compare the measured units of time. Thus, an average wind speed is calculated over time intervals Δt whose duration depends on the type of device used and using an N number of average values calculated over that interval. For calculating an hourly wind speed average, for example, if the average speed value is available at 5- or 10-min intervals, 12 values of the first type and 6 of the second will be used.

In practice, however, the average value of wind speed is calculated with Eq. (5), with a better approximation, for shorter the time intervals Δt:

$$v_m \cong \frac{\Delta t}{T} \sum_{j=1}^{\frac{T}{\Delta t}} v_j = \frac{1}{N} \sum_{j=1}^{N} v_j \; [m/s] \tag{5}$$

The net electric power at the output of the generator, which takes into account both the efficiency of the electrical part and the efficiency of the mechanical part of a wind group, is given by Eq. (6), according to Ref. [7]:

$$P_{el} = \frac{1}{2} C_e \rho A v_{m3}^3 \; [W] \tag{6}$$

where A is the swept area by the rotating wind turbine blades; C_e is the total net efficiency factor, which is determined at the terminals of the electric power transformer for the wind power group; ρ is the average air density at hub height; and v_{m3} is the average cube of the instantaneous wind speed.

On the other hand, due to the fact that the power of a wind power group is proportional to the wind speed at the third power, we can consider that the average cubic speed, defined as the root of the third order of the average cube of the instantaneous wind speed, according to Eq. (7), is a measure of the available power and energy for a wind power group:

$$v_{m3} = \left(\frac{1}{T} \int_0^T v_i^3(t)dt \right)^{\frac{1}{3}} \; [m/s] \tag{7}$$

The irregularity of the wind determines the difference between v_{m3} and v_m, so that the instantaneous wind speeds are all more different from their average value, and the more v_{m3} is greater than v_m. This pattern of the wind is characterized by the irregularity factor that is defined by Eq. (8).

$$k_f = v_m / v_{m3} \tag{8}$$

Often the irregularity factor can be replaced by the mean square deviation denoted as σ and expressed by Eq. (9):

$$\sigma \cong \left(\frac{1}{N} \sum_{j=1}^{N} \left(\frac{v_j}{v_m} - 1 \right)^2 \right)^{\frac{1}{2}} \tag{9}$$

The measurement units of the wind speed, with modifications of the wind division, are small intervals of 0.5 or 1 m/s; on this basis, they can be easily compared. For this purpose, the measured values of wind speed are classified into different speed classes. For each class, the probability of occurrence of measured values for wind speed allocated to this class is calculated with by taking into account the total number of measured values of wind speed. This frequency distribution always shows a typical course.

Mathematical approximations of such probability distributions can be performed with different functions that can be described by a small number of parameters. For the distribution of wind speeds, for example, either the Weibull distribution or the Rayleigh distribution can be used [7].

3. Wind energy production planning

Considering the assumptions made in the first section of this chapter about wind characteristics, we may note that having in advance information about the wind can be useful in the decision processes related to optimal power system operation. The process of obtaining mentioned information related in our case to the wind energy will be denoted as wind energy production forecasting.

Generally speaking, the forecasting term can be understood as the process of determining a sample P_{k+1} or a set of samples $\{P_{k+1+m} \mid m \in \mathbf{N}\}$ for a specific time t, given the set $\{C_{k-n} \mid n \in \mathbf{N}, n \leq k\}$, where C_{k-n} may consist of P_{k-n} measurements or of a more complex data.

Model class	Type of model	Remarks
Physical	Physical	Models based on meteorological parameters (temperature, atmospheric pressure, geographical and local conditions, environmental conditions, etc.)
Statistical	Auto-regressive (AR)	Approaches that rely on linear statistical models
	Autoregressive moving average (ARMA)	
	Autoregressive integrated moving average (ARIMA)	
Artificial Intelligence	Artificial neural network (ANN)	
	Fuzzy logic (FL)	
	Support vector machines (SVM)	

Table 1.
Classification of forecasting models.

Starting from this assumption, it can be seen that future data can be obtained starting from historical data.

Wind power forecasting respects the above definition, and the literature denotes it as direct forecasting approach. The need of more complex prediction methods to eliminate undesirable uncertainties requires a two-step approach to be adopted resulting in indirect methods for wind power forecast. First step consists in wind speed prediction followed by the usage of the turbine power curve for wind power determination [4]. For the former class of methods, the precision of the forecast is significantly influenced by the process of wind speed prediction [8], or if it is also considered the high degree of variability of wind-to-power curve, both terms play a key role in limited predictability of wind power generation [9].

Generally speaking, the prediction can offer short- or long-term predicted data that correspond to a short window period that contains one or more predicted points or a long-term interval, respectively. In case of wind power forecasting, due to the complexity of the problem, correlated with accuracy of the obtained results, short-term wind power forecast offers an appropriate solution for the task of optimal power system operation that covers but is not limited to power quality, power balance, or economic planning problems. For this purpose, a large number of forecasting models have been developed, which can be classified into three main classes: physical models, statistical models, and artificial intelligence technique-based models [6]. **Table 1** summarizes the classes of previous models.

In the next section, we will focus on exploring forecasting methods from the last class model. The approach was chosen starting from the ability of this type of algorithms that could offer proper results based on learnt patterns that are more appropriate than methods based on linear models.

4. Forecasting based on artificial intelligence

The obtaining of closest possible estimated values from the real ones is the main target. For these purposes, two types of neural networks are investigated, namely, recurrent artificial neural network (RANN) and long short-term memory (LSTM) networks. Their performances are evaluated through the mean absolute error

(MAE), mean absolute percentage error (MAPE), signed mean squared error (SMSE), and normalized mean squared error (NMSE) indexes.

4.1 Recurrent ANN

Feedforward network outputs are calculated based on the network input that is propagated from the input layer to one or more hidden layers and to the output layer through direct connections between the layers. Due to these connections, FFANNs are static networks. A neural network can have inverse connections, from an upper layer to a lower layer (e.g., from the output layer to the input layer), so the output of the network depends on inputs, outputs, and current, previous, and current state of the network, which gives dynamic behavior, and such a network is called a dynamic network [10]. Reverse connections are also called recurrent connections, hence the name of recurrent ANN (RANN).

Delays are introduced via reverse connections, so the response of the networks is influenced by the order in which the input vectors are presented. By this delay, information about the input data is stored, and the network may have different answers when, at the input, the same input vector is applied. This behavior makes it possible to approximate dynamic systems and presents an advantage in the field of forecasting [10].

The most used and known recurrent ANN topologies are as follows [11]:

- Jordan ANN (Jordan network or output-feedback recurrent ANN) is a feedforward network with a single-hidden layer and a context neuron for each neuron in the output layer (**Figure 1**). The purpose of the context neuron is to maintain the activation (output) of a neuron in the output layer at time k until it is used at time $k + 1$. The connections between the output neurons and the context neurons are weighted, as are the direct connections [11]. In Refs. [10, 12], delay blocks are used, with the same purpose as context neurons, when moments k, $k + 1$, $k + 2$, ... are moments of time.

- ANN Elman (Elman network or globally recurrent ANN) uses the same context neurons or delay blocks; the difference with Jordan networks is that each hidden layer will have a layer made up of context neurons that are

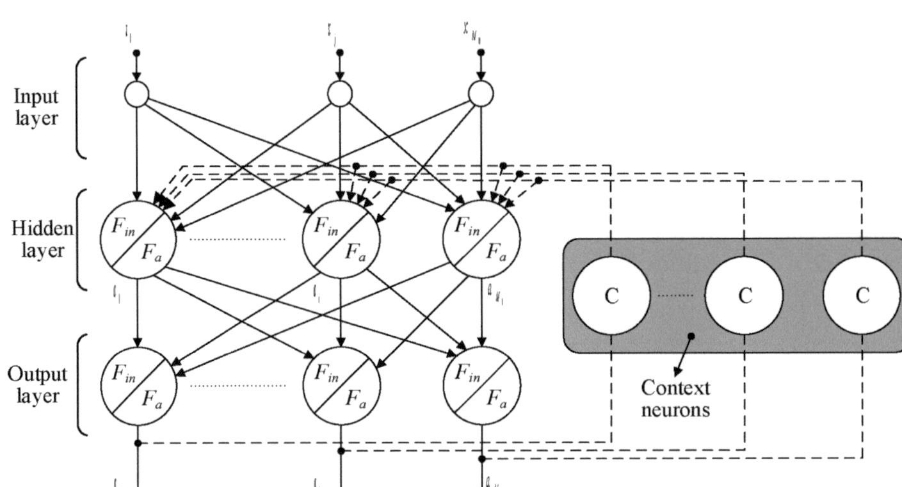

Figure 1.
Structure of a Jordan type RANN [11].

connected, further on, to inputs of neurons from the hidden layer (**Figure 2**) [10, 12].

- Completely recurrent ANN, each neuron in a hidden layer or output layer has one or more context neurons, so more information is retained [12].

The networks shown in the previously mentioned figures are global recurrent networks, where each context neuron connects to the input of each neuron in the hidden layer. If each context neuron links only to the input of the neuron to which it is assigned, the network is locally recurrent [12].

Recurrent ANNs can be trained using the generalized delta learning rule [10]. The weights of the connections between the layers and the displacement weights have a direct effect and an indirect effect on the activation of neurons. The direct effect is created by the weights of the connections between the layers, which can be calculated with the generalized delta rule.

The indirect effect is created by the weights of the connections between the context neurons and the neurons in the hidden layer to which they are connected. The inputs of a layer, which come from the outputs of the context neurons, depend on the same weights on which the outputs of the neurons connected to the context neurons depend. For this reason, the calculation of the gradient

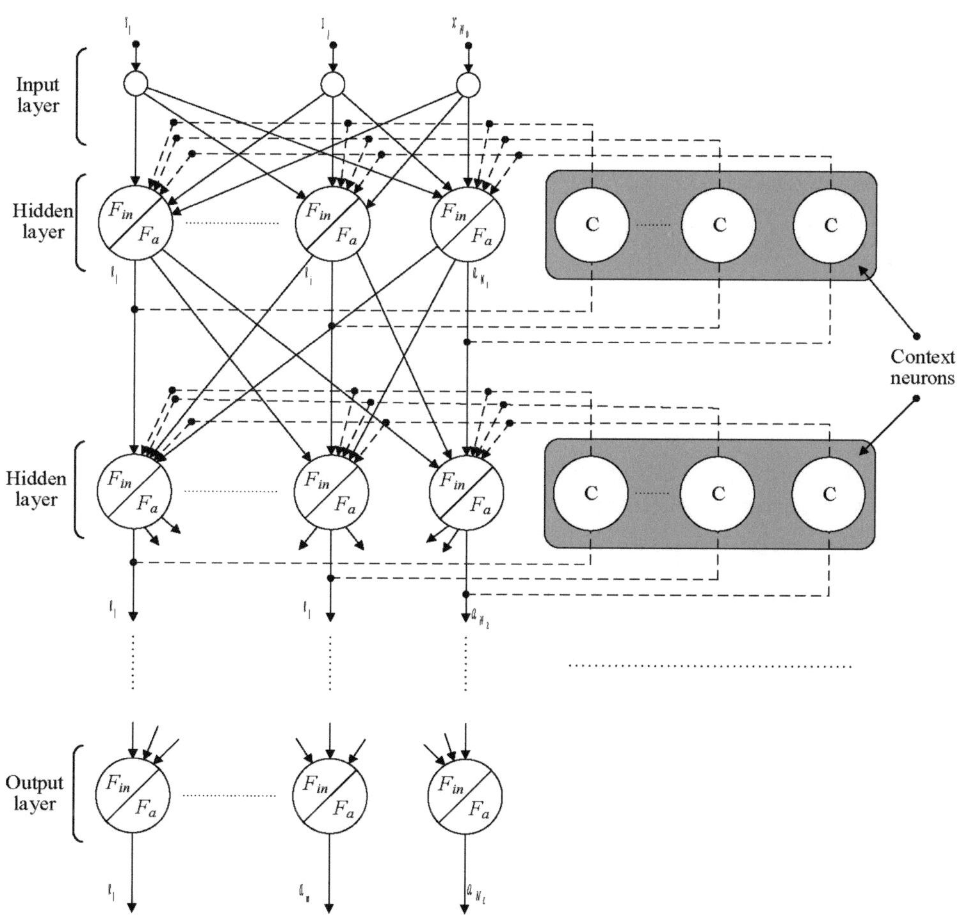

Figure 2.
Structure of the recurrent Elman type network [11].

depends not only on the weights of the network but also on the previous outputs of the network [12].

There are two different processes for calculating the gradient: the time-propagated generalized delta rule (back propagation through time—BPTT) and real-time recurrent learning (RTRL) [10]. In the first method, the gradient is calculated from the last time moment to the first time moment. For this reason, it is necessary to calculate the network response for each time point before calculating the gradient. In the second method, the gradient is calculated at each time point, together with the network response at that time, continuing with the rest of the time points [10]. The difference between the two methods is that the BPTT algorithm performs offline training and requires less computing power than the RTRL training algorithm that performs online training but requires greater computing power [10, 12].

The two methods are detailed in Ref. [10], and briefly, the steps of the two methods are as follows:

- The neural network is initialized, as in the case of the binary preceptor training algorithm and the generalized delta rule. In addition, the RTRL method requires the initialization of the previous values corresponding to the network delays.

- The network response is calculated. For the RTRL method, the response for the first time point is calculated, and for the BPTT method, the network response is calculated for each time point.

- Calculate the total derivatives that take into account the indirect and direct effects and the explicit derivatives that only take into account the direct effects. In the case of the RTRL method, these calculations are repeated for each time point, and in the case of the BPTT method, it is calculated at the last time point, starting from the last time point and continuing until the first time point.

- Calculate the derivatives of the error function. Using the results, the weights are updated, and the algorithm of training the binary preceptor and the generalized delta rule is continued.

Recurrent networks, through inverse connections and dynamic behavior, have a more complex error surface than static feedforward networks. This complexity is due to the nonlinear behavior of the error function, and it has several local minima. Also, a small change in the weights can lead to significant changes in the error in increasing direction [12].

The descending gradient method uses, depending on the network parameters, the partial first-order derivative of the error function, so it is a first-order learning algorithm. When the partial derivative of the second order is used, additional information is obtained on the gradient, and the methods that use this information are called second-order algorithms [12]. Some of these methods are: Newton's method, the conjugated gradient method, and the scaled conjugate gradient method. The last two methods are detailed and described in Ref. [12], and Newton's method is described in Ref. [10].

The main disadvantage of the recurring ANN is given by the inverse connections of the recurring networks. They may have a delay order greater than the first order to store several previous network states. Due to the fact that the value of the gradient may depend on previous values, for a high delay order, the value of the

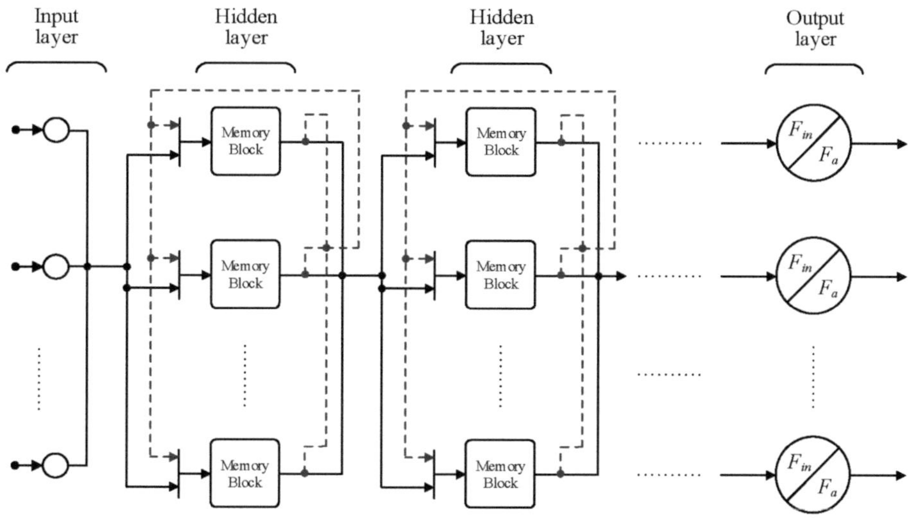

Figure 3.
The schematic diagram of an LSTM network [13].

gradient may drop very rapidly to an infinitesimal value (vanishing gradient) or increase to a large value (exploding gradient) [13].

4.2 Recurrent ANN of LSTM type

Long short-term memory (LSTM) networks are recurrent networks that have the ability to memorize/learn short-term dependencies but for a long time. It solves the problem of the vanishing gradient by maintaining the local error at a constant value or in a certain domain, so the value of the gradient does not reach infinitesimal values or very large values [13, 14].

Compared to recurrent networks that have neurons in the hidden layer and context neurons or connections with delay blocks, LSTM networks have blocks of memory in the hidden layer. Each memory block contains one or more memory cells, an input gate, an output gate, and, optionally, a forget gate [13]. The schematic diagram of an LSTM network and of a memory block is shown in **Figure 3**.

The role of the cell is to maintain and transmit information from the input of the memory block to the output. The input gate determines the information that enters the cell, and the output gate detects the information coming out of the memory block. The gates control this by calculating the weighted amount of the gate entries and the weight of each entry. This sum is sent to a unipolar sigmoid function, where a value between 0 and 1 is obtained, which controls what information enters the cell and what information exits the memory block [14].

The inputs of the memory block are propagated forward to the input gate, forget gate, and exit. Each gate and circle containing the symbol $\sum$ and a block representation of the unipolar sigmoid function or hyperbolic tangent in **Figure 4** are represented as artificial neurons because the mathematical operations that are applied to the inputs are identical and simplify the graphical representation.

In addition to the entry weights: $\omega_g, \omega_i, \omega_f,$ and $\omega_o,$ there are also moving weights: $b_g, b_i, b_f,$ and b_o.

To determine the output of the memory block [14], the current state of the cell is calculated before it is affected by the input gate and the forget gate, denoted by g in **Figure 4**:

$$g = F_g(y_g) = F_g\left(\sum_{j=1}^{m} \omega_{gj} \cdot x_{gj} + b_g\right) \tag{10}$$

where $i = \overline{1, m}$, and m represents the number of inputs. After this, the output of the input port i is calculated:

$$i = F_i(y_i) = F_i\left(\sum_{j=1}^{m} \omega_{ji} \cdot x_{ji} + b_j\right) \tag{11}$$

The output of the forget gate u can be determined as:

$$u = F_u(y_u) = F_u\left(\sum_{j=1}^{m} \omega_{ju} \cdot x_{ju} + b_j\right) \tag{12}$$

These three values, g, i, and u, together with the previous state of the cell $c^{<k-1>}$, are used in determining the current state of c cell:

$$c = u \cdot c^{<k-1>} + i \cdot g \tag{13}$$

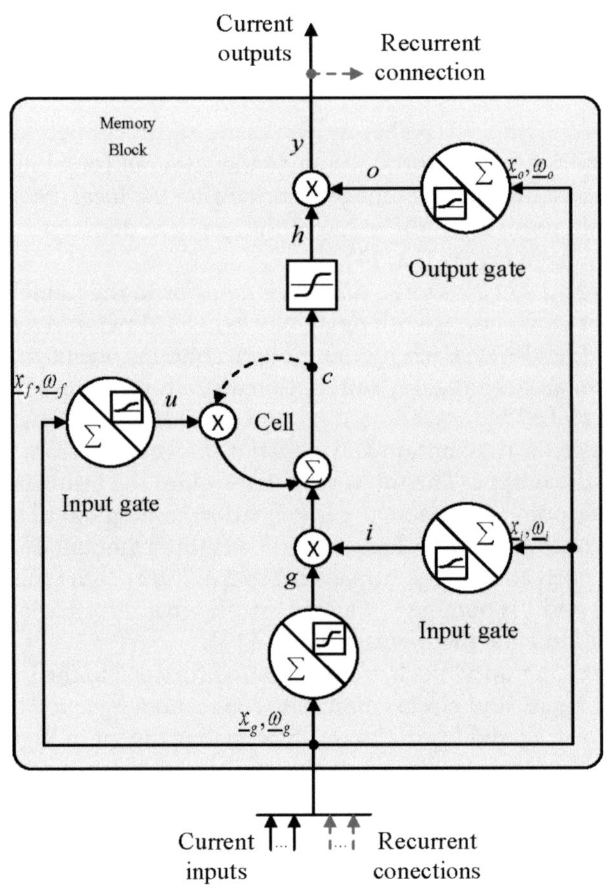

Figure 4.
Principle diagram of a memory block (dotted line connections are recursive connections that have first-order delay) [13].

Further, the state of the cell evolves through the hyperbolic tangent function F_h, and h is determined. It is further calculated the activation of the output gate o, after which the two are multiplied to obtain the output of block y:

$$o = F_o(y_o) = F_o\left(\sum_{j=1}^{m} \omega_{jo} \cdot x_{jo} + b_j\right); y = o \cdot F_h(c) \tag{14}$$

The LSTM networks are trained using a slightly modified version of the BPTT algorithm and of the RTRL algorithm. For the output gate, RTRL is used, and for the rest of the gates and elements of the memory block, RTRL is used. The modification consists in the fact that the errors are considered only for updating the weights of one block, without the error of the other block being modified. The effect created by the gates is that the error can only pass unchanged through the cells. The training algorithm is detailed in Ref. [13].

Following is the implementation of the models discussed for the prediction and the results obtained.

5. Analysis of the neural network-based forecasting models

5.1 RNN-based forecasting model

The chosen RNN network has one input layer, one hidden layer, and one output layer. The hidden layer has recurrent connections from exits to inputs, being a recurrent Elman type network (**Figure 2**).

The activation function of hidden layer neurons is the unipolar sigmoid function, which is why the time series is normalized in the domain [0 1], and the activation function of the output neurons is the linear function.

In addition to the weights of feedforward connections, the recurrent network contains weights for the recurrent links. These are initialized with random values in the domain [−0.1 0.1], and the random values have a uniform distribution (white noise). The weights of feedforward connections are initialized with the value 0, and the displacement weights are initialized with the value 1.

For the network training and the forecasting, the walk-forward method is used, and the parameter assignment is performed by the experimental method. The error statistics for different values of the network parameters are presented below.

By using the graphs shown in **Figures 5–8**, an RNN model was chosen with the following parameters: a window size of one sample, a single neuron in the

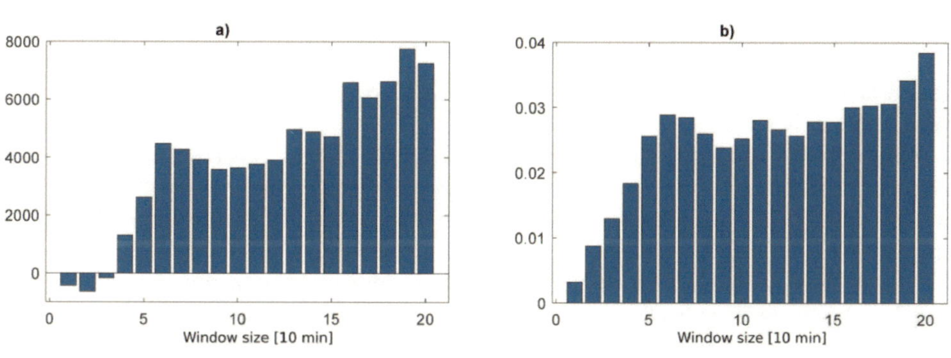

Figure 5.
Window size. (a) SMSE, and (b) NMSE.

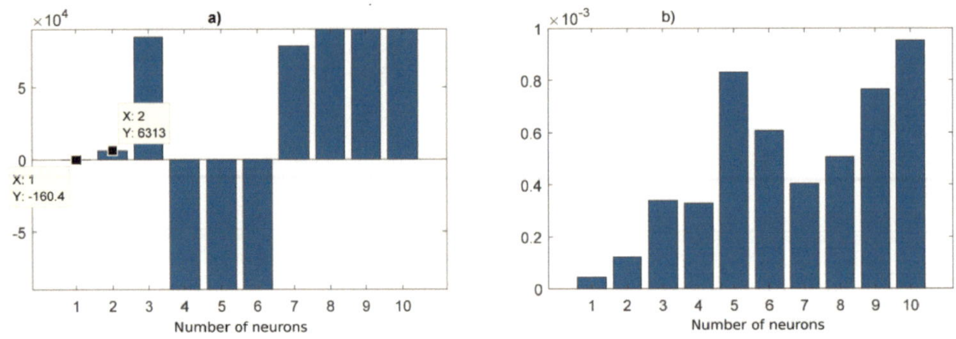

Figure 6.
Error statistics for neuron number from the hidden layer. (a) SMSE, and (b) Theil's U statistic.

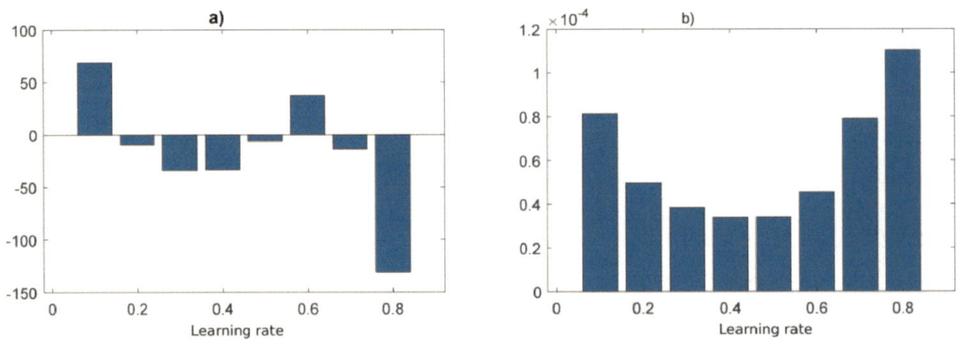

Figure 7.
Error statistics for learning rate. (a) SMSE, and (b) Theil's U statistic.

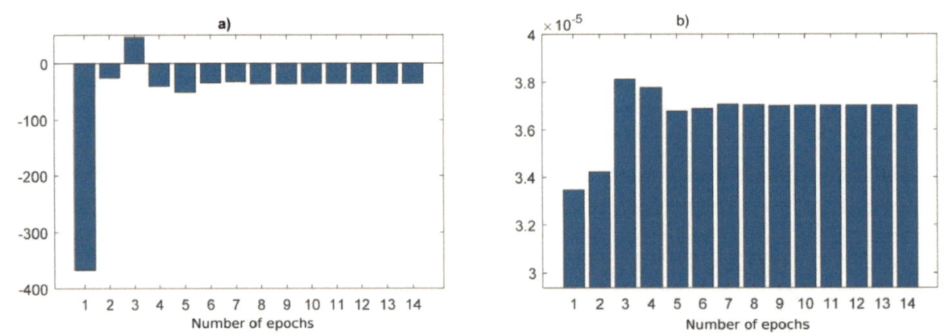

Figure 8.
Error statistics for number of training epochs for a window. (a) SMSE, and (b) Theil's U statistic.

hidden layer containing the recurrent links, the learning rate of 0.4, and two-training periods.

Although the window is one sample, the number of previous steps is four samples. In addition to these parameters, tests were performed for the moment of rate learning (0.2), the slowing of the learning rate (0), and the number of previous samples (4).

The forecast realized with the RNN (4-1-1) model and the performance of the model is shown in **Figures 9** and **10**.

From the above figures and **Table 2**, it is observed that the RNN model succeeds in a good forecast of the time series when the forecast horizon is a sample ahead, the maximum error being 408.353 MW, which means 15.2% of the maximum electricity.

Next, the aim is to identify the impact of the horizon prediction increasing. To increase the prediction horizon, six samples were considered before corresponding to 1 h. The same iterations of the parameters were repeated, and the chosen model is RNN (6-13-6) with the following parameters: a window size of one sample, 13 neurons in the hidden layer containing the recurrent links, a learning rate of 0.4, the moment of learning rate 0.5, the slowing of the learning rate is omitted (0), and two periods of training.

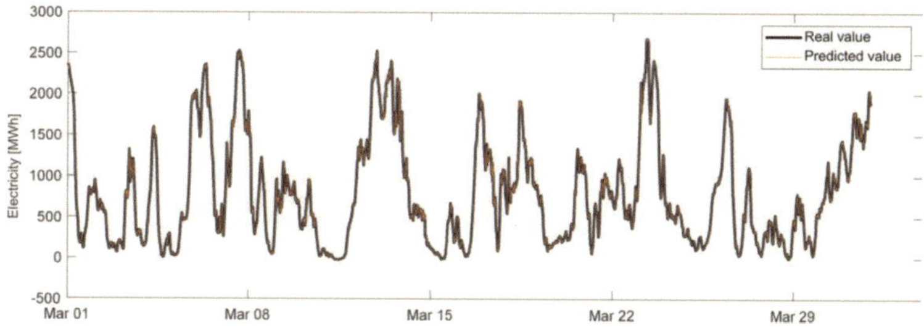

Figure 9.
Forecast performed with the model RNN (4-1-1).

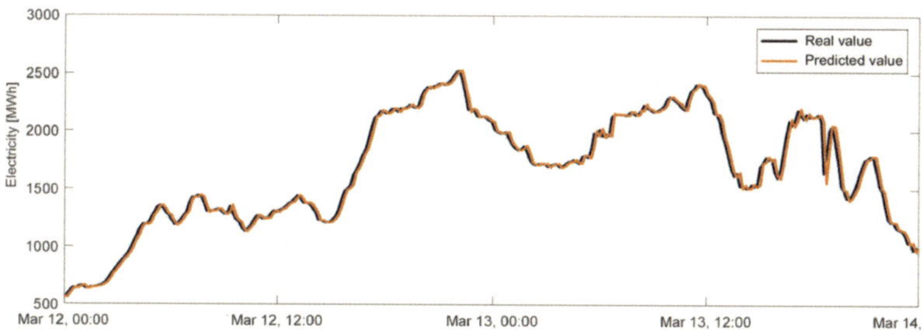

Figure 10.
Forecast performed with the model RNN (4-1-1), from March 12 to 14.

MAE	MAPE	SMSE	NMSE	Theil's U statistics
23.02	5.66	−25.61	0.0030	3.42×10^{-5}

Table 2.
RNN (4-1-1) model performances.

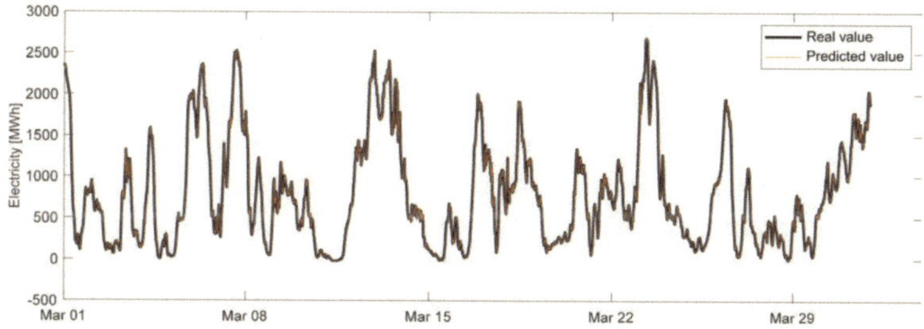

Figure 11.
Forecast performed with the model RNN (6-13-6).

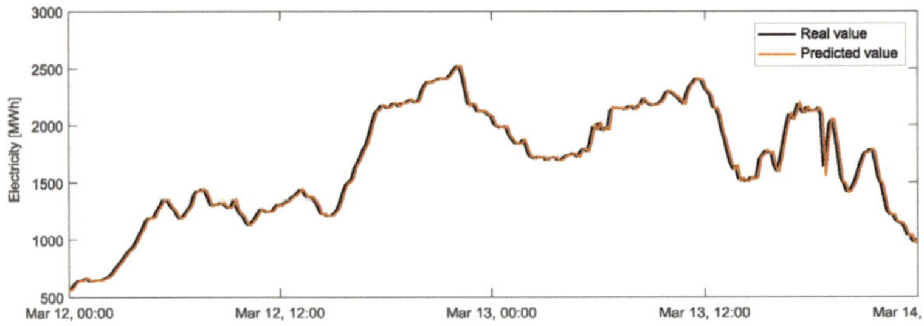

Figure 12.
Forecast performed with the model RNN (6-13-6), from March 12 to 14.

MAE	MAPE	SMSE	NMSE	Theil's *U* statistics
21.35	5.27	−9.28	0.0028	3.27×10^{-5}

Table 3.
RNN (6-13-6) model performances.

Although the window is one sample, the number of previous steps is six samples.

It can be observed that at every six samples or window moving, a forecast of the next six samples is made.

The performance and forecast achieved with the RNN model (6-13-6) for a forecast horizon of six samples are presented in **Figure 11**.

From **Figure 12**, it can be seen that the RNN model manages to approximate the time series, the performance being synthesized in **Table 3**.

Although the forecast horizon was widened by five samples, performances similar to those obtained for the single-sample prediction were obtained, with the difference that the model requires more neurons in the hidden layer.

5.2 LSTM implementation and analysis of the model based on LSTM networks

The LSTM model is similar to the RNAFF and RNN model; it contains an input layer, a hidden layer, and an output layer. **Figure 3** represents the schematic of the model, with the difference that there is only one hidden layer.

Instead of neurons, the LSTM network contains memory blocks that have inputs, outputs, and different gates (input, output, and forget). For the inputs and outputs, the hyperbolic tangent transfer function is used, and for the gates, an estimate of the unipolar sigmoid activation function is used. Since the inputs and outputs of a memory block use the hyperbolic tangent function, the time series is normalized in the interval $[-1\ 1]$.

For the network training and the forecasting, the walk-forward method is used, and the parameter assignment is performed by the experimental method. The error statistics for different values of the network parameters are determined further.

By using the graphs shown in **Figures 13–16**, a LSTM model was chosen with the following parameters: a window size of one sample, five neurons in the hidden layer containing the recurrent links, a learning rate of 0.4, and two-training periods. Although the window is one sample, the number of previous steps is three samples. In addition to these parameters, tests were performed for the moment of learning rate (0.3) and the slowing of the learning rate (0).

The forecast realized with the LSTM model (3-5-1) and the performance of the model is described below (**Figures 17** and **18** and **Table 4**).

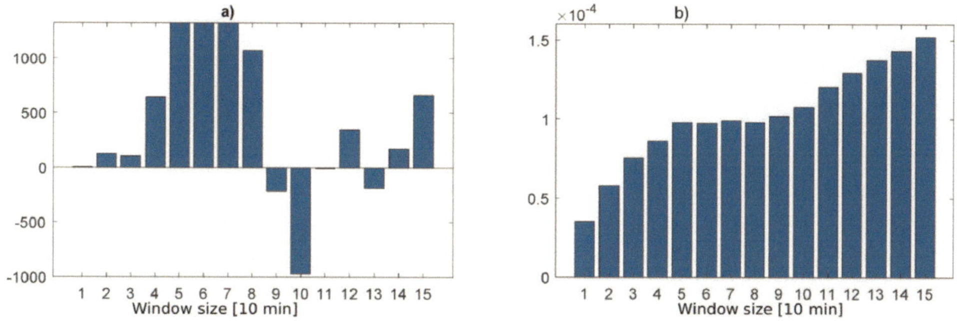

Figure 13.
Statistics for window size. (a) SMSE, and (b) Theil's U statistic.

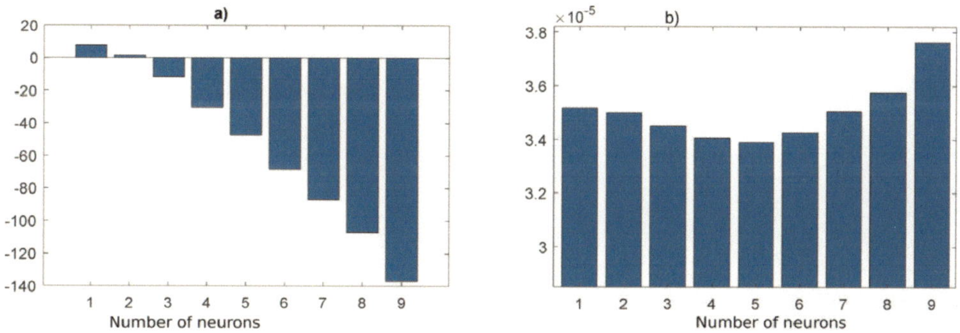

Figure 14.
Statistics for neuron number from the hidden layer. (a) SMSE, and (b) Theil's U statistic.

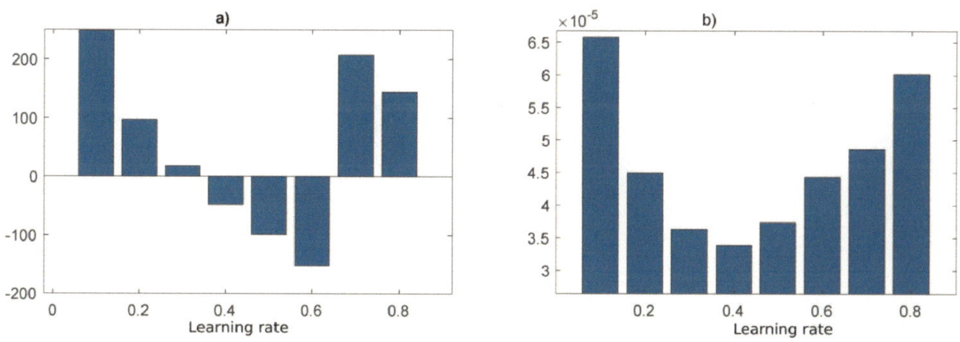

Figure 15.
Statistics for learning rate. (a) SMSE, and (b) Theil's U statistic.

The LSTM model has performances almost identical to the RNN model, the maximum error being 408,441 MW meaning 15.21% of the maximum electricity.

Similar to the previous models, to increase the prediction horizon, six samples were considered before corresponding to 1 h. The same iterations of the parameters were repeated, and the chosen model is LSTM (1-29-6) with the following parameters: a window size of one sample, 29 neurons in the hidden layer containing the recurrent links, the learning rate of 0.6, the moment of learning rate 0.5, the slowing of the learning rate is omitted (0), and two training times, and only one previous sample is used. Similar to the RNN model, at every six samples or window movement, a forecast of the next six samples is made.

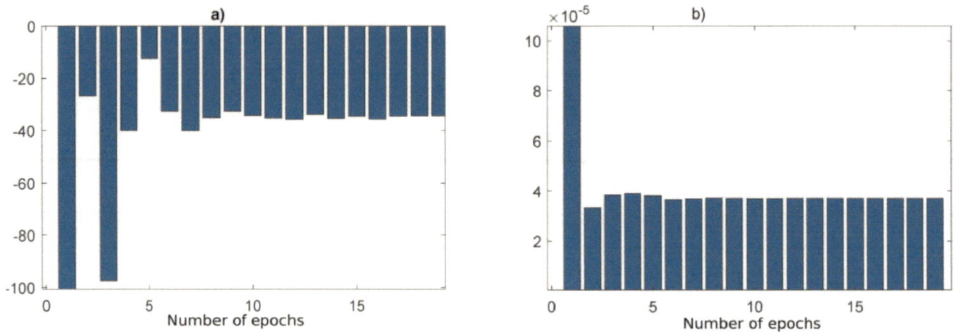

Figure 16.
Statistics for number of training epochs for a window. (a) SMSE, and (b) Theil's U statistic.

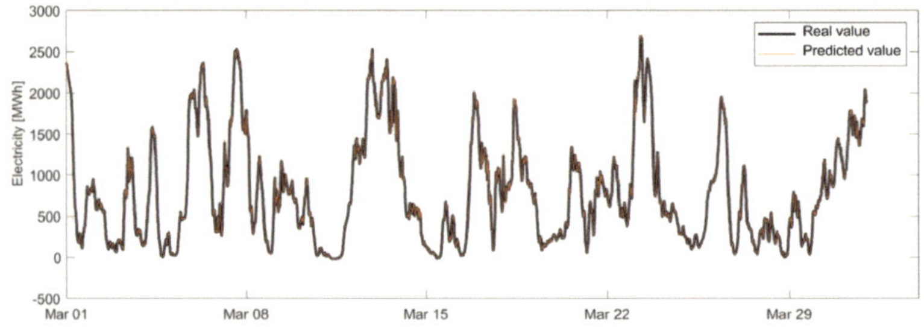

Figure 17.
Forecast performed with the model LSTM (3-5-1).

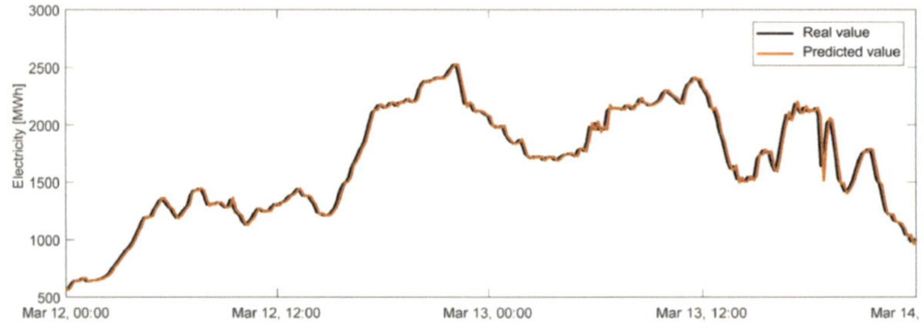

Figure 18.
Forecast performed with the model LSTM (3-5-1), from March 12 to 14.

MAE	MAPE	SMSE	NMSE	Theil's *U* statistics
22.17	5.42	−26.82	0.0029	3.33×10^{-5}

Table 4.
LSTM (3-5-1) model performances.

The performance and forecast achieved with the LSTM model (1-29-6) for a six-sample forecast horizon are presented below (**Figure 19** and **Table 5**).

The performances are similar to the RNN and LSTM models with a sample prediction horizon and the RNN model with six forward samples. The maximum error is 391.435 MW, which corresponds to 14.57% of the maximum electricity.

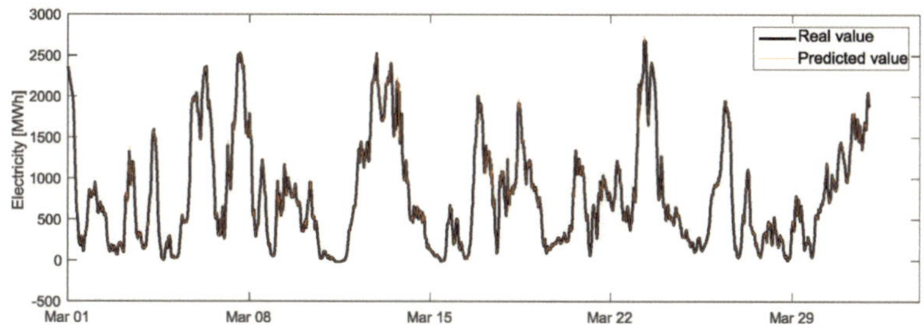

Figure 19.
Forecast performed with the model LSTM (1-29-6).

MAE	MAPE	SMSE	NMSE	Theil's *U* statistics
21.36	5.22	−41.33	0.0027	3.25×10^{-5}

Table 5.
LSTM (1-29-6) model performances.

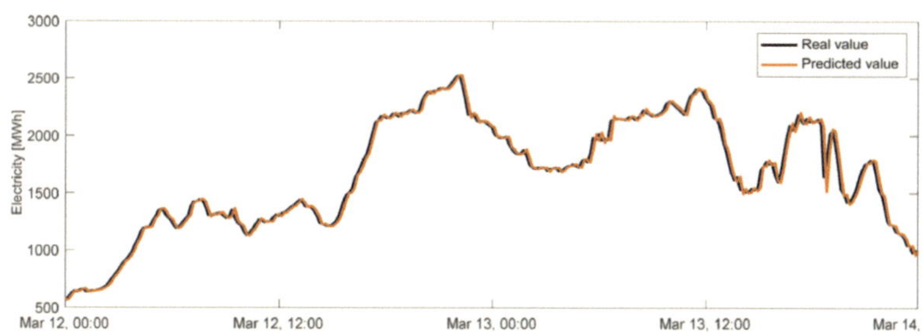

Figure 20.
Forecast performed with the model LSTM (1-29-6), from March 12 to 14.

Model	MAE	MAPE	SMSE	NMSE	Theil's *U* statistics
RNN (4-1-1)	23.02	5.66	−25.61	0.0030	3.42×10^{-5}
LSTM (3-5-1)	22.17	5.42	−26.82	0.0029	3.33×10^{-5}

Table 6.
Models performances for a horizon of one sample.

5.3 Models comparison

5.3.1 Models comparison for a prediction horizon of one sample (10 min)

The compared models are RNN (4-1-1) and LSTM (3-5-1). The used comparison period is 1 day (March 13), and the performances are determined on the entire time series.

Between the RNN and LSTM models, the differences are very small as can be seen in **Figure 20** and **Table 6**, the LSTM model being with MAE—3.7%, MAPE—4.24%, NMSE—3.33%, and Theil's *U*—2.63% more efficient than the RNN model, but the SMSE statistics is 4.72% higher.

For the forecast horizon of a sample and the time series used, the RNN and LSTM models present the best performances (**Figure 21**).

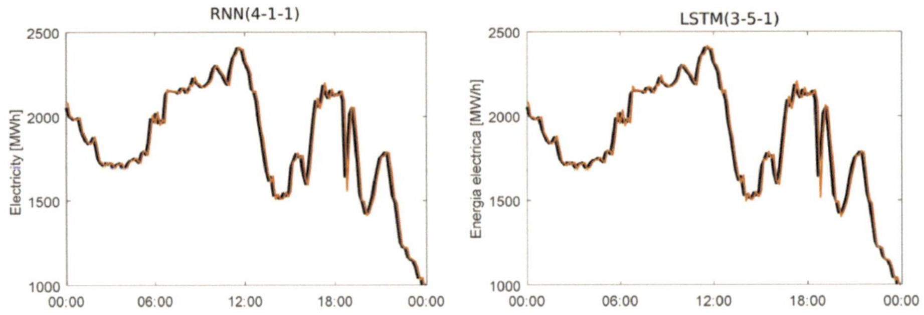

Figure 21.
Comparison of the two models in achieving the forecast with a horizon of one sample.

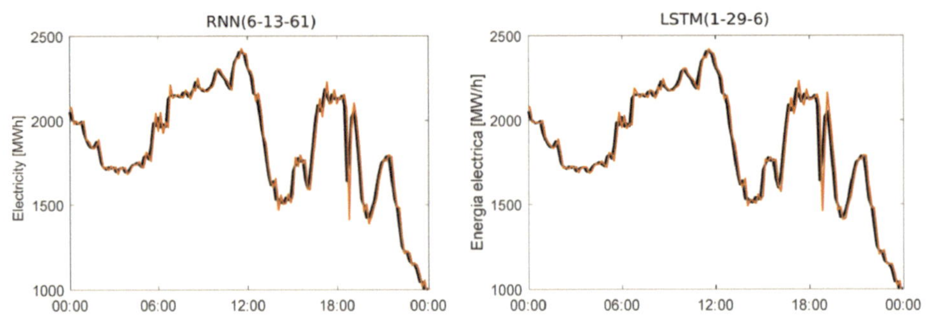

Figure 22.
Comparison of the two models in achieving the forecast with a horizon of six sample.

Model	MAE	MAPE	SMSE	NMSE	Theil's *U* statistics
RNN (6-13-6)	21.35	5.27	−9.28	0.0028	3.27×10^{-5}
LSTM (1-29-6)	21.36	5.22	−41.33	0.0027	3.25×10^{-5}

Table 7.
Models performances for a horizon of six samples.

5.3.2 Performance comparison for a prediction horizon of six samples (1 h)

The compared models are RNN (6-13-6) and LSTM (1-29-6). The period used for comparison is the same, and 1 day (March 13) and the performances are determined for the entire time series (**Figure 22** and **Table 7**).

Models based on recurrent networks manage a better approximation of the time series and achieve similar or even better performance than the one-sample prediction before. Between RNN and LSTM, the differences are small between MAE, MAPE, NMSE, and Theil's *U* statistics, and the largest difference appears in SMSE statistics. It is worth mentioning that the LSTM model has performances similar to the RNN model and has a single sample before and the same size as the window of a sample.

5.3.3 Performances comparison of the RNN and LSTM models for different values of the forecast horizon

The increasing forecast horizon leads inevitably to higher errors and lower performance. For the forecast, the same models were used: RNN (6-13-6) and

LSTM (1-29-6), and the forecast horizon ranges from 6 to 144 samples with a step of six samples between tests.

From the figure above, it can be observed that the two models have similar performances up to a prediction horizon of 50 samples, after which the difference between the models increases, the performances being better for the RNN model.

The forecasts made with the two models for a prediction horizon of 144 samples (1 day) are presented below.

In **Figure 23**, a significant prediction error is observed for the first forecast made, after which the next forecast has a much lower prediction error. Although the models are trained on each window, the forecast is only performed on multiples of 144 samples. This simplifies the arrangement of the resulting values, each of the 144 future samples being saved in a vector that is added to the previous prediction vector. To eliminate this initial error, more frequent predictions can be made, even at each window (10 min in this time series), but manipulating the results is more difficult.

From **Figures 23** and **24**, it can be observed that the two models have similar values for the small variation zones of the time series and for the larger variation

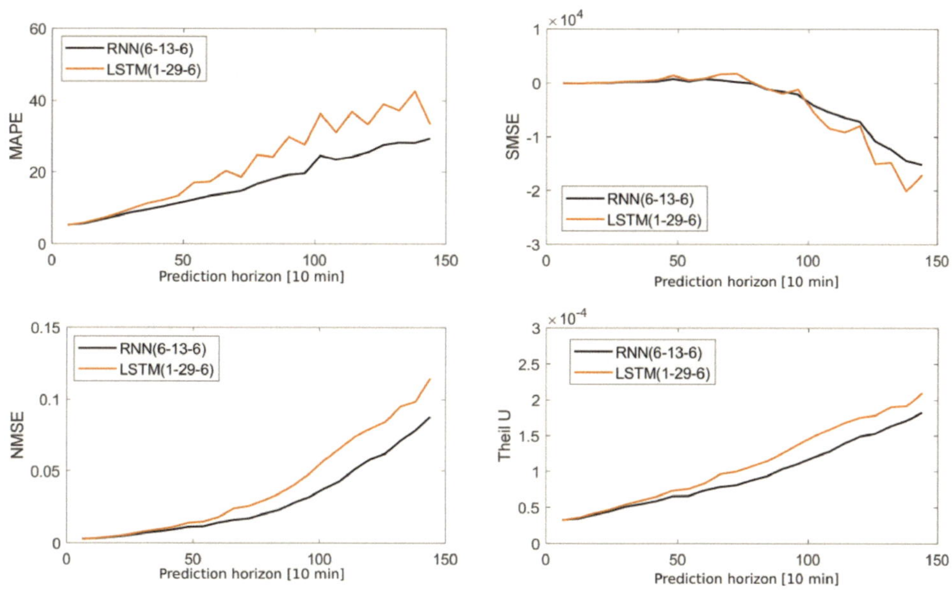

Figure 23.
Models RNN (6-13-6) and LSTM (1-29-6) comparison for different values of the prediction horizon.

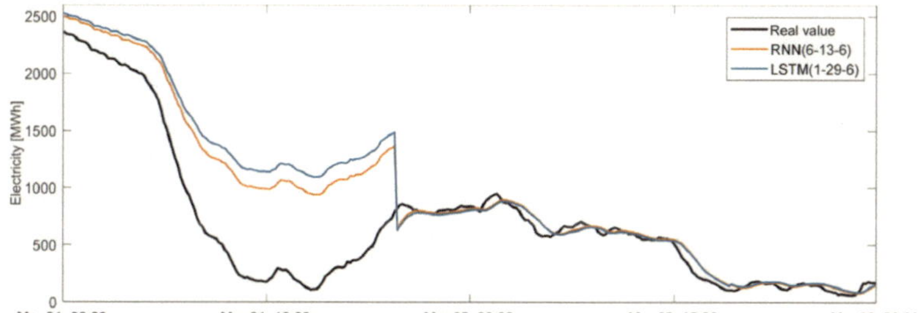

Figure 24.
Models RNN (6-13-6) and LSTM (1-29-6) comparison and a 144 samples horizon, March 1 to 3.

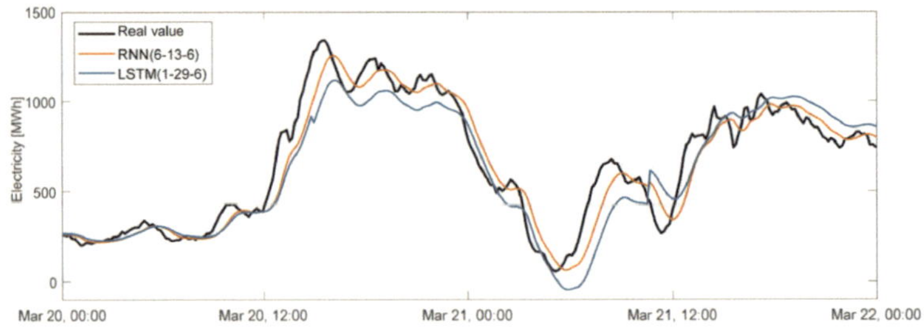

Figure 25.
Models RNN (6-13-6) and LSTM (1-29-6) comparison and a 144 samples horizon, March 20 to 22.

Model	MAE	MAPE	SMSE	NMSE	Theil's *U* statistics
RNN (6-13-6)	111.86	29.47	−1514.3	0.0881	1.83×10^{-4}
LSTM (1-29-6)	127.98	33.40	−1708.3	0.1147	2.10×10^{-4}

Table 8.
RNN (6-13-6) and LSTM (1-29-6) models performances comparisons for a 144 samples prediction horizon.

zones, the values differ, the more efficient model being the RNN model (**Figure 25** and **Table 8**).

The parameters of the models have not been reconfigured, most likely, if a longer training (more epochs), a larger window, and a number of previous samples are allowed, the model's performances would increase.

For the chosen parameters of the models and the time series taken into account, the RNN model presents the best performances for a prediction horizon of 144 samples (1 day). For a better performance, it may be considered the reconfiguration of parameters, the networks structure changing such as adding more hidden layers, the activation function changing, or the usage of another type of neural network.

6. Conclusions

In this chapter, two prediction models based on artificial neural network (recurrent type neural networks: RNN and LSTM) for wind power forecasting were studied. Experimental results exhibited were obtained through the data processing performed with the aim of Python platform with Keras library and Tensorflow.

It can be noted that there are a multitude of models recorded in literature, and some are generalized and can be applied to multiple domains, and others are more specific to one domain or application. Furthermore, no other variables, except the production of electrical energy, have been considered. For example, the wind speed and weather forecasts can be used to forecast, and results with and without these added variables can be compared.

To achieve the proposed objective of forecasting, only one variable was used, namely, historical records of electricity production from wind energy, but several variables that influence this size can be considered.

In addition to highlighted problems, other directions of research consist in identifying stochastic models and models based on neural networks, capable to approximate the seasonal and trend components of time series. There is possible to adopt models that reduce the number of parameters to be assigned, such as the model presented in Ref. [15].

The RNN and LSTM models were compared for different time horizons. For the chosen parameters, the RNN model presents the best performances in the case of the short-term horizon, the higher the forecast horizon, the greater the difference between the RNN and LSTM model results. It is worth mentioning that, the performances of the models depend on the chosen parameters, for this reason, different methods of parameter optimization can be engaged, resulting new methods that can further be investigated.

Author details

Cristian-Dragos Dumitru* and Adrian Gligor
George Emil Palade University of Medicine, Pharmacy, Science and Technology of Targu Mures, Targu Mures, Romania

*Address all correspondence to: cristian.dumitru@umfst.ro

References

[1] Eurostat. Renewable energy statistics. Data extracted in January 2019 [Internet]. 2019. Available from: https://ec.europa.eu/eurostat/statistics-explained/index.php/Renewable_energy_statistics#Wind_power_becomes_the_most_important_renewable_source_of_electricity [Accessed: 01 September 2019]

[2] Wagner H-J, Mathur J. Introduction to Wind Energy Systems. Green Energy and Technology: Springer International Publishing AG; 2018. 105 p. DOI: 10.1007/978-3-319-68804-6_1

[3] Gilbert C, Browell J, McMillan D. Leveraging turbine-level data for improved probabilistic wind power forecasting. IEEE Transactions on Sustainable Energy. 2019:1-9, forthcoming. DOI: 10.1109/TSTE.2019.2920085

[4] Wang Y, Hu Q, Srinivasan D, Wang Z. Wind power curve modeling and wind power forecasting with inconsistent data. IEEE Transactions on Sustainable Energy. 2018;**10**(1):16-25

[5] Vatră CA. Renewable sources productivity—Particulary case wind turbine. In: Lucrările Conferinței Naționale şi Expoziția de Energetică CNEE 2007; 7-9 November 2007; Sinaia: Romania; 2007. pp. 917-922

[6] Gasch R, Twele J, editors. Wind Power Plants: Fundamentals, Design, Construction and Operation. Heidelberg, Dordrecht, London, New York: Springer; 2012. 548 p. DOI: 10.1007/978-3-642-22938-1

[7] Golovanov N, Postolache P, Toader C. Eficiența şi calitatea energiei electrice. Bucureşti, Romania: AGIR; 2007

[8] Zhang Y, Wang J, Wang X. Review on probabilistic forecasting of wind power generation. Renewable and Sustainable Energy Reviews. 2014;**32**: 255-270. DOI: 10.1016/j.rser.2014.01.033

[9] Naik J, Dash S, Dash PK, Bisoi R. Short term wind power forecasting using hybrid variational mode decomposition and multi-kernel regularized pseudo inverse neural network. Renewable Energy. 2018;**118**: 180-212. DOI: 10.1016/j.renene.2017.10.111

[10] Hagan MT, Demuth HB, Beale MH, De Jesús O. Neural Network Design. Stillwater, Oklahoma: Martin Hagan; 2014

[11] Kriesel D. A Brief Introduction to Neural Networks, 27 May 2005. [Internet]. 2019. Available from: http://www.dkriesel.com/en/science/neural_networks [Accessed: 29 April 2018]

[12] Medsker LR, Jain LC. Recurrent Neural Networks: Design and Applications. Boca Raton, FL: CRC Press; 2001

[13] Gers F. Long Short-Term Memory in Recurrent Neural Networks

[14] Olah C. Understanding LSTM Networks, 27 August 2015. [Internet]. Available from: http://colah.github.io/posts/2015-08-Understanding-LSTMs/ [Accessed: 23 May 2019]

[15] Ghiassi M, Saidane H, Zimbra DK. A dynamic artificial neural network model for forecasting time series events. International Journal of Forecasting. 2005;**21**:341-362

Chapter 2

A Research Framework for the Multidisciplinary Design and Optimization of Wind Turbines

Luca Sartori, Stefano Cacciola, Alessandro Croce and Carlo Emanuele Dionigi Riboldi

Abstract

The design of very large wind turbines is a complex task which requires the development of dedicated tools and techniques. In this chapter, we present a system-level design procedure based on the combination of multi-body numerical models of the turbine and a multilevel optimization scheme. The overall design aims at the minimization of the cost of energy (COE) through the optimization of all the characteristics of the turbine, and the procedure automatically manages all the simulations required to compute relevant loads and displacements. This unique setup allows the designer to conduct trade-off studies in a highly realistic virtual environment and is an ideal test bench for advanced research studies in which it is important to assess the economic impact of specific design choices. Examples of such studies include the impact of stall-induced vibrations on fatigue, the development of active/passive control laws for large rotors, and the complete definition of 10–20 MW reference turbines.

Keywords: wind energy, design, optimization, MDO, system design

1. Introduction

Wind energy has known a rapid transformation in recent years, and the increasing use of wind power as a strategic asset has supported a continuous growth in the size of wind turbines. This is mainly due to the fact that larger turbines allow a higher energy production and a faster recovery of the investment [1]; however, progressive upscaling requires technological developments in order to successfully manage the analysis and design of very large rotors. In fact, modern wind turbines exhibit an intrinsic multidisciplinary behavior which makes it difficult to disentangle the effects of different design aspects. Complexity arises from a variety of phenomena: for example, an increasing length of the beam-like elements implies a high flexibility of the rotor blades and the tower, which originates high displacements and deformations when the turbine is loaded. This produces a highly nonlinear and continuous interaction between the aerodynamic field and the underlying structure, so that the dynamic behavior of the turbine is typically fully coupled. When the turbine is deployed offshore, the motion of waves, currents, and tides contributes to further complicate the inflow conditions and the response of the system. In addition, the extreme flexibility of the turbine components means that the eigenfrequencies of the turbine are relatively low, which increases the risk of dangerous superposition among different turbine modes or between the natural frequencies of the turbine and the *per-revolution* ones introduced by the rotating motion. The consequences of unresolved resonances can be perceived through the dynamic response of the system, which could cause a sharp increase of the fatigue damage along the turbine, but also increased vibration problems and instabilities. In this scenario, unstable behaviors can also be triggered by purely aeroelastic effects, like flutter- or stall-induced vibrations [2, 3].

To address these issues, a novel approach is needed to successfully conduct the design of next-generation wind turbines. In fact, most current design strategies rely on a sequential process where different aspects of the project (i.e., aerodynamic design, structural verification, control tuning) are conducted separately according to discipline-specific procedures and methodologies. As the complexity of the phenomena increases, however, it's easy to see how such an approach is prone to some limitations: in a coupled system where each part of the design has a cascade effect on all other components it would be hard, if at all possible, to separate the scope into a number of subproblems which are then treated individually. A modern approach, on the contrary, must forcibly be based on the integration of different expertises and procedures coming from different disciplines into a unified perspective at system level.

In recent years, there has been a great effort from the wind energy community to develop such a multidisciplinary vision, and this required to renegotiate the assumptions of most design applications and, in particular, a redefinition of what are the fundamental design drivers. While classic approaches usually optimized the turbine to produce the maximum energy production, it is now widely accepted that modern turbine design should aim at the minimization of the cost of energy (COE) [4–7]. This is often achieved by defining the COE as the merit function of a dedicated multidisciplinary design optimization (MDO) algorithm. The advantage of targeting the COE in lieu of individual figures like the annual energy production (AEP) or the blade mass is that the computation of the cost of energy is directly affected by technical, managerial, and economical characteristics of the project which are condensed into a unified parameter. This way, it is much easier to properly evaluate the impact of certain design choices on the overall performance of the system. The effective implementation of an MDO procedure, however, is not a trivial task as multiple choices and solutions are possible, and typically, every design framework supports a different philosophy.

A first, fundamental aspect to discuss is the computation of the cost of energy: at present, most MDO algorithms available in the literature rely on the COE as the objective function to drive their optimization process. However, the way this figure is calculated can dramatically influence the optimization and potentially drive the optimization in completely different directions.

Another important aspect to leverage on concerns the physics of the turbine: the numerical modeling of wind turbines has improved significantly, and different models of increasing complexity are available. While the choice of low- or high-fidelity models certainly has some impact, it is absolutely necessary that the models are able to capture the multidisciplinary aspects of the wind turbine behavior. Usually, higher-fidelity models intrinsically support a better physical description, but this comes at the expenses of the computational time.

A third fundamental choice is the domain of optimization in terms of how many design variables are included into the multidisciplinary optimization algorithm. To support a system-level design, in fact, different unknowns coming from different disciplines should be collected together. These parameters ideally account for the blade shape, the internal structure of the blades and tower, and the general features of the rotor and, potentially, of the wind farm. Usually, combining a satisfactory number of variables to a sufficient quality of the physical modeling is hard and often compromises have to be made. The nature of such compromises and the direction they favor typically characterize a certain approach to the MDO design.

1.1 State of the art in the multidisciplinary design of wind turbines

Numerical optimization methods have been successfully applied to specific wind turbine design tasks, in particular the shape optimization of the rotor and the structural sizing of blades and tower. Recently, several authors have proposed different approaches toward the implementation of multidisciplinary design algorithms. These contributions have been quite heterogeneous and show wide differences in the methods, the level of modeling, and the definition of the design variables. In the following, we provide a brief survey of the most prominent contributions in the field of MDO design of wind turbines.

In this context, a broad family of MDO applications has its focal point on the design of the rotor, and a great deal of contributions is available in the literature. For example, Kenway and Martins [8] were among the first to conduct a truly coupled aero-structural design, in which variables related to chord, twist, airfoils, and structure were optimized to minimize the COE at a certain wind site. Later, Xudong et al. [9] introduced an aeroelastic model of the wind turbine and optimized the aerodynamic shape of the rotor under limits on the axial thrust and the shaft torque. In this work, the structure was indirectly optimized through the thickness of the blade. More recently, Pourrajabian et al. [10] applied a genetic algorithm to design a small-scale wind turbine rotor. Here the design variables account for chord and twist as well as the shell thickness, while the design objectives include the starting time and the power output. Bottasso et al. [11] developed a procedure in which the chord and twist of the blade are optimized together with the thickness of all the structural components. Here, the procedure is based on a multistage process in which the rotor shape is firstly optimized and the resulting solution is then fed to a structural optimization problem. A peculiarity of this application is that the structural optimization is based on a broad set of fully resolved design load cases (DLC). The problem of defining an accurate yet computationally affordable load basis has been investigated in detail by several authors. For example, Sessarego and Shen [12] proposed an optimization method based on surrogate modeling which include a very large set of load cases. In this case, the chord, twist, and thickness of a 5 MW rotor are successfully optimized, together with the spar caps and the shear webs, for a set of hundreds of load cases. An alternative methodology has been proposed by Pavese et al. [13] and is based on the substitution of computationally expensive turbulent cases with custom-made shear zones and equivalent deterministic loads. This way, the authors were able to embed the load computation directly within the workflow of the MDO algorithm and to re-optimize a 10 MW rotor with the reduced load basis. Several authors have applied genetic algorithms to the design of rotors: Zhu et al. [14], for example, combined the blade element momentum (BEM) theory to a three-dimensional finite elements method (FEM) to conduct the aero-structural optimization of a 1.5 MW rotor blade. Aerodynamic design variables in this work include the rotor speed and the chord distribution, while structural ones account for the thickness and the width of the spar caps and the positioning of the shear webs. The optimization is then managed by a non-dominated sorting genetic algorithm. A similar study in which BEM and finite elements are combined has been conducted by Dal Monte et al. [15] although the focus here is on small-scale rotors.

Very recently, Yang et al. [16] proposed a similar methodology to redesign a 2.1 MW rotor. A particular aspect of this work is that airfoil shapes are described by means of Taylor series and directly designed as part of the optimization problem. The possibility of designing the airfoils together with other aero-structural variables is a prominent topic and will be extensively discussed in this work. A recent application of rotor-based MDOs concerns the study of solutions for very large rotors and, possibly, load-mitigating techniques to support lightweight ones. In this light, for example, Richards et al. [17] proposed an integrated aeroelastic design of a 100 m blade. The study considers the impact of bond-line damage, a common failure mode for long blades, on the design and explores how an MDO-based design can reduce its influence on the integrity of the wind turbines. Pavese et al. [18], instead, employed an MDO approach to optimize the sweep of the blade achieving an important load alleviation on a 10 MW rotor.

More recently, several researchers extended the concept of multidisciplinary design to the whole wind turbine, introducing the idea that accurate modeling of physics, coupled to suitable optimization schemes, could pave the way for a system-engineering approach to the design of modern wind turbines. An early realization of this perspective was proposed by Fuglsang et al. [19], who first combined aeroelastic simulations to an optimization cycle based on the COE. Here, the design variables are some macro-parameters of the wind turbine like rotor shape and diameter, tower height, and rotor speed, whose impact on the COE is computed through a dedicated model. Later on, Maki et al. [20] discussed an MDO based on a multilevel design architecture. In this framework, an external optimization is employed to minimize the cost of energy. At this level, design variables include rated power, rotor diameter, rotor speed, and other turbine parameters. This loop is interfaced with two sublevels, which conduct discipline-based optimizations. The first allows to maximize the AEP of the rotor, while the second minimizes the loads produced by the blades.

Another formulation at system-level has been presented by Ashuri et al. [21, 22]. Here, a preliminary round of design can be made by optimizing blade length, tower height, and the rotor speed for minimum COE. Then the optimal configuration is fed to a second optimization problem which manages aero-structural design variables from the rotor and the tower. The design unknowns, which are simultaneously designed together, account for the chord, the twist, the blade structure, and the tower. The simulation environment is built around the solver FAST, which allows to take full DLCs into considerations during the design.

Significant research on MDO algorithms has been done at NREL, where Dykes et al. [23] developed an integrated tool for the analysis and design of a wind energy project from component level to the entire wind plant. The algorithm is based on the OpenMDAO software by Gray et al. [24] which manages the optimization problems and links the different models required for the various simulations, supporting increasing levels of detail depending on the needs. The engine of the algorithm is the internal cost models which recursively computes the COE by taking information from the various modules. A similar project, also based on OpenMDAO, has been recently developed by Zahle et al. [25]. The structure of the information flow basically allows to conduct a full aero-structural optimization of large wind turbines, and the simulation environment is widely based on internally developed software. The tool includes some particular specialisms, like the possibility to design the airfoils during the optimization and a frequency-based fatigue model which reduces the computational time.

1.2 Overview of this research

In this chapter, we present a detailed approach to MDO optimization of a wind turbine, which is the result of more than a decade of research and development in the field.

The fundamental idea put forward is to create a design framework which combines a detailed multi-body modeling of the wind turbine and a specifically engineered optimization structure, which relies on the combination of different design layers. Within this architecture, the complex multidisciplinary design problem is conducted by various modules which individually perform the optimization of specific aspects of the system including the aerodynamic design, prebend optimization, tower and structural design, and the automatic synthesis of the control laws. An additional outer loop manages the overall optimization of the wind turbine, thus ensuring that the design modules work harmoniously toward the minimization of the cost of energy.

This approach is a novelty in wind energy, since previous research was based either on monolithic approaches or on multilevel design but, usually, with only low-fidelity models or a limited load spectra. One of the essential advantages of this construction is the possibility to maintain a possibly large number of DLCs directly within the design, so that optimal solutions automatically comply with international certification standards. Another important feature of this modular scheme is that it makes possible to introduce new modules and procedures without the need to redesign the whole software. Due to its intrinsic generality and flexibility, this approach provides an ideal platform to conduct cost-oriented design studies.

2. Multilevel MDO optimization of wind turbines

As discussed, a modern approach to wind turbine design should implement an MDO approach and identify a certain balance between conflicting requirements of scope, number of design variables, modeling accuracy, and computational time. The development of our design tool Cp-Max was initiated by Bottasso et al. [26] and continuously expanded through subsequent research activities by Gualdoni [27], Bortolotti et al. [28], and recently by Sartori [29]. The main feature of the integrated MDO approach here proposed is the nested/multilevel architecture shown in **Figure 1**. The idea is to interface two different layers of design, one dedicated to the preliminary optimization of the turbine and the other based on as a series of individual design modules which perform detailed design of specific wind turbine subcomponents. The main advantage of such construction is that it allows to combine the level of detail of highly physical simulations with a high number of design variables typical of aero-structural design problems. Another important feature is that different design variables, which form a highly heterogeneous set, are not treated all at the same level, like it happens in monolithic approaches. Vice versa, the variables are divided among the different design modules, so that quantities affecting a certain merit function could be grouped together and provide a better sensitivity to the optimization scheme. A fundamental assumption behind this architecture is that the external loop *and* all the modules work together in the same direction, that is, the target of each design step must be coordinated to the others, in order to avoid a scenario where the design within a module evolves independently from that at the global/macro level. This is summarized in **Figure 1** where different modules act on different contributions of the merit function, whereas the outermost layer acts on the minimization of the COE as a whole. During the optimization, all the required simulations are automatically managed and performed by our multi-body simulation tool Cp-Lambda (see Bottasso and Croce [30] for details).

2.1 Macro design loop

The macro design loop (MDL) is the outer layer of Cp-Max, and it drives the global optimization of the wind turbine, as well as the data flow between the various

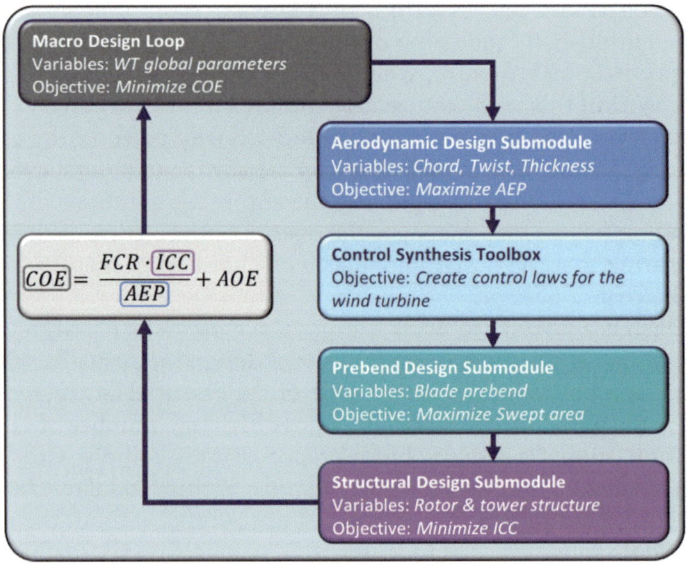

Figure 1.
Architecture of the MDO algorithm implemented in Cp-Max.

submodules. For a certain target of power rating and wind class, the main task of the MDL is to conduct a preliminary design in which few important (macro) parameters of the turbine are optimized in order to minimize the cost of energy according to the well-known scaling model from Fingersh et al. [31]. However, given the multilevel nature of Cp-Max, for each perturbation of those parameters, one or more submodules at the inner level perform the detailed design of the desired aspects of the wind turbine. This can be formalized by the following optimization problem:

$$\left(\boldsymbol{p}_a^*, \boldsymbol{p}_b^*, \boldsymbol{p}_s^*, \boldsymbol{p}_g^*, \boldsymbol{r}_c^*, COE^*\right) = \min_{\boldsymbol{p}_g} \left(COE\left(\boldsymbol{p}_a, \boldsymbol{p}_b, \boldsymbol{p}_s, \boldsymbol{p}_g\right)\right) \tag{1}$$

$$\boldsymbol{p}_g = \left[R, h_{Hub}, \theta_\pi, \gamma_b, \sigma_c^g, \tau_c^g, \sigma_t^g, \tau_t^g\right] \tag{2}$$

$$s.t. : \boldsymbol{g}_g\left(\boldsymbol{p}_a, \boldsymbol{p}_b, \boldsymbol{p}_s, \boldsymbol{p}_g\right) \leq \boldsymbol{0} \tag{3}$$

Eq. (1) shows that the MDL takes as input all the design variables subdivided into different sub-arrays: $\boldsymbol{p}_a$ identifies aerodynamic design variables, $\boldsymbol{p}_b$ is the set of prebend design variables, $\boldsymbol{p}_s$ are the structural ones, and $\boldsymbol{p}_g$ identifies the macro design variables. The latter are directly optimized by the MDL, while the other families are progressively optimized by the various submodules. The set of the control laws $\boldsymbol{r}_c^*$ is also produced as part of the optimal solution. It must be noticed that in this notation, a star superscript identifies optimal quantities. The array of the macro design variables includes a set of global features of the wind turbine. As shown in Eq. (2), these include the rotor radius R, the nominal hub height h_{Hub}, the rotor tilt angle θ_π, the blades collective coning angle γ_b, and four additional shape parameters $\sigma_c^g, \tau_c^g, \sigma_t^g, \tau_t^g$.

The role of these parameters is to provide a channel to connect the MDL with the individual design submodules and in particular to the aerodynamic design. These parameters are computed for the rotor blade starting from the spanwise distribution of chord and thickness as follows:

$$\sigma_c^g = \frac{3A_b}{A_\pi} = \frac{3\int_0^R c(r)dr}{\pi R^2} \tag{4}$$

$$\tau_c^g = \frac{\int_0^R rc(r)dr}{A_b} \tag{5}$$

$$\sigma_t^g = \frac{1}{100}\int_0^1 t(\eta)d\eta \tag{6}$$

$$\tau_t^g = \frac{\int_0^1 \eta t(\eta)d\eta}{\int_0^1 t(\eta)d\eta} \tag{7}$$

where A_b is the blade solid area, A_π is the ideal rotor area, $c = c(r)$ is the blade chord distribution as a function of the local radius r, and $t = t(\eta)$ is the distribution of blade percent thickness as a function of the nondimensional length coordinate η. The detailed design of the blade shape, in terms of chord, twist, and percent thickness is not managed at the MDL level: in fact, this would require a certain number of dedicated design variables which should be added to the array $\boldsymbol{p}_g$, with consequences on the computational time. Instead, the blade shape is actually designed by the aerodynamic submodule, which performs a detailed sizing for the maximum AEP. The module performs a genuinely aerodynamic design, which means that, if the problem is not properly constrained, the evolution of the aerodynamic design is completely independent from the underlying structure, which is sized at a later stage. Within an economic perspective, the risks of such an approach are clear: it might happen that the aerodynamic design evolves toward a highly efficient blade with low solidity and reduced sectional thickness. This would require to design a heavier structure and, ultimately, would achieve a higher cost of energy. To avoid such kind of bad positioning of the MDL optimization, the four shape parameters are used to constrain the aerodynamic design within certain limits. Those limits, in turn, are directly managed by the MDL on behalf of the aerodynamic design module. Eq. (3) includes all the nonlinear inequality constraints acting on the MDL optimization problem. Although a full global design is typically unconstrained, it is possible in Cp-Max to introduce constraints on the maximum individual loads so that it is possible, for example, to find the optimal design for a certain load spectrum.

2.2 Aerodynamic design submodule

The aerodynamic design submodule (ADS) performs the optimization of chord, twist, and thickness of the blades in order to maximize the AEP of the rotor. During this stage, the internal structure of the blades is frozen, so that it is possible to avoid the computation of the dynamic loads and deflections required to redesign the structure. However, this strategy is reasonable only when the blade shape is not changed significantly during the aerodynamic design. Otherwise, if the blade plan-form or its thickness is broadly modified during the optimization, this would typically require a complete redesign of the structure which might potentially overshadow the advantages of an improved energy capture. Then, it is vital to ensure that the external MDL loop has enough authority on the individual design modules to avoid that the outcomes of a single design step march against the global strategy of minimizing the COE. In the ADS, this is done by applying the four shape parameters defined as part of the MDL design variables as nonlinear constraints of the aerodynamic optimization problem. As a consequence, the aerodynamic submodule conducts a constrained maximization of the AEP for given target values of those parameters. This automatically assures that the variations are not too sharp during the design. Following the notation introduced by Eqs. (1)–(3), the workflow of the ADS can be summarized in the following optimization problem:

$$(\boldsymbol{p}_a^*, AEP^*) = \max_{\boldsymbol{p}_a} \left(AEP\left(\boldsymbol{p}_a, \boldsymbol{p}_b, \boldsymbol{p}_s, \boldsymbol{p}_g\right)\right) \tag{8}$$

$$\boldsymbol{p}_a = \left[\boldsymbol{p}_{a_c}\, \boldsymbol{p}_{a_\theta}\, \boldsymbol{p}_{a_t}\right] \tag{9}$$

$$s.t.: v_{tip} \leq v_{tip_{max}} \tag{10}$$

$$\boldsymbol{g}_a\left(\boldsymbol{p}_a, \boldsymbol{p}_g\right) \leq \boldsymbol{0} \tag{11}$$

Eq. (8) suggests that the ADS provides the optimization of the aerodynamic variables; however, given that all the required simulation are based on a complete multi-body model of the rotor, it is possible to account for cross-disciplinary effects like the deformation of the blades at this step. This explains why all design variables and not only those directly pertaining to the ADS enter the module. Eq. (9) shows the composition of the aerodynamic design variables: in particular, the array $\boldsymbol{p}_a$ encompasses parameters related to the chord ($\boldsymbol{p}_{a_c}$), twist ($\boldsymbol{p}_{a_\theta}$), and nondimensional thickness ($\mathbf{p}_{a_t}$) of the blade. To limit the number of degrees of freedom, all these distributions are controlled by specific parameterizations based on additive or multiplicative gains (see Gualdoni [27] for details).

At the beginning of each evaluation, the aerodynamic model of Cp-Lambda is updated with the actualized values of the design variables $\mathbf{p}_a$. Once the model is correctly updated, $C_p - TSR - \beta$ curves are computed as a three-dimensional surface for varying values of the collective pitch angle β and tip-speed ratio TSR. From the envelope of the $C_p - TSR - \beta$ curves, it is possible to compute the ideal regulation trajectory for the wind turbine, under the assumption that the control strategy is based on the tracking of the optimal power C_p^* when the turbine works in the partial-power region. A constant power strategy is instead considered for the full-power range. It should be noticed that, according to Eq. (10), the automatic buildup of the power curve must ensure at any time that the maximum tip speed is lower than a certain bound, in order to constraint the acoustic emissions of the rotor. Once the theoretical power curve is known, it is weighted by the site-specific Weibull distribution to assess the AEP. Eq. (11) includes a list of nonlinear inequality constraints acting on the aerodynamic design problems. As mentioned, four constraints are always enforced by the MDL through the four shape parameters of Eqs. (4)–(7). The possible additional constraints include maximum values for chord and twist as well as maximum gradient of the chord, twist, and thickness distributions.

2.3 Control synthesis tool

Looking at the many DLCs accounted for in the design of a novel wind turbine, it can be noticed that the control system plays a relevant part in many of them. For this reason, a realistic control system is an unmissable complement to the aeroelastic simulation tool in the ideal flowchart of an MDO procedure.

Based on the multilayer paradigm proposed by Bottasso et al. [32], considering horizontal axis wind turbines, the functions of a controller can be arranged into decoupled layers, each targeting a control objective by means of a suitable control law, designed independently from the others.

The innermost layer implements a trimming control, tasked with piloting the turbine so as to produce the design power output corresponding to the current wind. The outcome of this layer is typically a set point for collective pitch and generator torque. The measurements needed for this part are most basically the rotational speed of the rotor (or electrical generator) and the wind itself. The latter is usually measured for supervisory tasks, i.e., in order to feed a gain scheduler, accounting for the different behavior of the turbine in partial-power, transition, and full-power regions.

The actual trimming law design can be chosen based on several considerations. Of course, the simplest PI laws make for a generally easier-to-design alternative, thanks to the lower number of control gains (see [33, 34]). Control laws of this type are not model-dependent and do not need any detailed knowledge of the features of the machine. This in turn increases robustness with respect to potential deficiencies in the reduced model used for control design.

As possible drawbacks, tuning procedures are usually heuristics without an assured performance, and some form of manual tweaking is needed to increase performance to the desired level. Furthermore, unless semi-automatable numerical algorithms for gain synthesis are deployed which would make a possible adoption of these laws for more complex tasks [34], in order to reduce the effort necessary for control tuning, these controllers are typically purely single-input single-output. This limitation can be dealt with in a smart way by analyzing the operative spectrum of the turbine. Trimming can be obtained in the partial-power and transition regions by governing torque only through a PI law based on rotational speed, setting pitch at a predefined set-point function of the wind speed. In the full-power region, torque is kept at a preassigned value, whereas pitch is changed according to a PI law, again based on a measurement of the rotational speed [33]. Another control scheme operates with a PI loop closed on the rotational speed all over the operational spectrum for pitch, and torque assigned using a pre-determined schedule with respect to the wind speed [34].

Alternatively, more sophisticated model-based trimming control laws can be adopted. These are based on a reduced model, assembled from state equations for all the dynamics which are required for an accurate description of the interaction between machine states and controls (see [11]). Typically, for trimming purposes it is necessary to include state equations for torque balance, as well as for pitch and torque actuator dynamics. Clearly, if the number of state variables is greater than 1 and controls are simultaneously collective pitch and generator torque, the ensuing law is implicitly multiple-input multiple-output. Very reliable methods for the computation of optimal gains can be deployed to automatically compute the gain matrix for this type of control problems, with some guarantees on control performance obtained a priori under design conditions.

Model-based trimming laws are lighter to manage than model-independent laws, do not make distinction between regions in the operating envelope, and are suitable for extensions, obtained by considering individually the pitch input of each blade and including states corresponding to deformable systems on board, like the tower, shaft, and blades (in so doing, incorporating in the trimming layer also functions typical to the outer control layers). Of course, model-based control is more exposed to robustness issues in case of inaccuracies in the reduced model. Furthermore, in order to supply the control laws with all necessary measurement signals, observers for deformative states may be needed (see [35], Chapter 5).

It has been shown in [36] how to simply extend the basic feedback capabilities of a model-based trimming control law to take advantage of simple LiDAR measurements, providing a reading of the average wind speed measured in a volume ahead of the turbine. This is exploited to better cope with fast changes in the average speed of the stream, thus helping reducing machine downtime and ultimate loads.

In any case, due to the change in the behavior of the turbine over the operative regions, it is often necessary to include a form of model or gain scheduling in trimming control laws, in turn requiring the adoption of wind measurements or more stable and accurate wind observers (see [37–39]).

Concerning outer layers, the frequency band of pitch actuators can be employed to target the lower per-revolution frequencies showing up in the spectra of load signals from the blades, tower, and shaft. Works by Bossanyi [33] propose an approach based on Coleman's transformation, developed to target higher harmonics in the works of Cacciola [40–42], whereas Bottasso et al. [32] describe a model-based approach for deterministic loads and a model-free approach for residual noise components, typically due to high-frequency turbulence. Synthesizing the gains of this control layer, whatever the selected approach, can be done according to heuristics or optimal numerical procedures [43].

As stated at the beginning of this section, in the flow of an MDO problem, a trimming controller is needed for all load cases where the machine is not parked, and load-targeting layers may be included usually to the aim of making the machine lighter.

During an iteration, when the aerodynamics and structure of the model have been assigned, all elements needed to synthesize a control law are available. The analysis of $C_P - TSR - \beta$ curves, considered together with required rated power and limitations on the blade peripheral speed for complying with noise constraints, produce an optimal regulation trajectory [11]. This is readily translated into set points for the rotational speed, input collective pitch, and electrical torque for increasing values of the wind between cut-in and cut-out speed.

For model-independent trimmers, these equilibrium points are the only needed information. Gains usually designed a priori, and typically scheduled with respect to the wind speed, can be applied online to trim the machine to the desired set point. Empirical methods for slightly changing these gains, in view of modifications in the model brought in by the evolution of the turbine design solution over the optimization, may be accounted for. A redesign of the PI gains iteration by iteration, by means of dedicated optimal numerical procedures, is usually not possible in the realm of an optimization, unless based on dynamic simulations carried out on a reduced model, thus imposing a lower computational effort than when run on the high-fidelity aeroelastic code.

In this sense, model-dependent trimmers offer a good level of simplification. Once the regulation trajectory has been obtained, it is possible to readily synthesize a reduced model, linearized around selected operative conditions along the trajectory. Gains are then synthesized for every considered equilibrium condition. When going online, previously computed gains are simply scheduled with respect to the wind following a linear interpolation scheme.

As said, it is possible to include more states in the reduced model than those related to dynamic rotational equilibrium. Most typically, a decoupled equation for the tower fore-aft dynamics, obtained from the truncation to the first mode of a modal representation of the tower top displacement, can be included with good results on tower base loads and basically no effect on trimming performance. In the workflow of the MDO, once a high-fidelity model has been prepared, a model reduction can be carried out, obtaining the coefficients to feed the corresponding equation of the reduced model. Considering Cp-Lambda, it can be reported that a modal reduction tool can be found in the code, suitable for the task.

DOI: http://dx.doi.org/10.5772/intechopen.90172

Both in case deformative states are accounted for or not, the presence of a reduced model allows to consider more accurately the evolution of the optimal turbine design, iteration by iteration.

2.4 Prebend design submodule

In those designs where the maximum tip displacement is an active constraint, the prebend design submodule (PDS) is used for optimizing the prebend, i.e., the native out-of-plane deflection of the blade. During the optimization loop, the PDS tries to design the prebend so that, when the rotor undergoes normal loading, the resulting swept area is maximized. It must be noticed that this criterion can be translated into the requirement of minimizing the parameter A_δ as defined by Eq. (12):

$$A_\delta = \int_0^1 \delta_y(\eta) d\eta \tag{12}$$

where $\delta_y(\eta)$ is the spanwise distance between the deformed blade and the ideal rotor plane. This way, the parameter A_δ corresponds to the gray-shaded area in **Figure 2**.

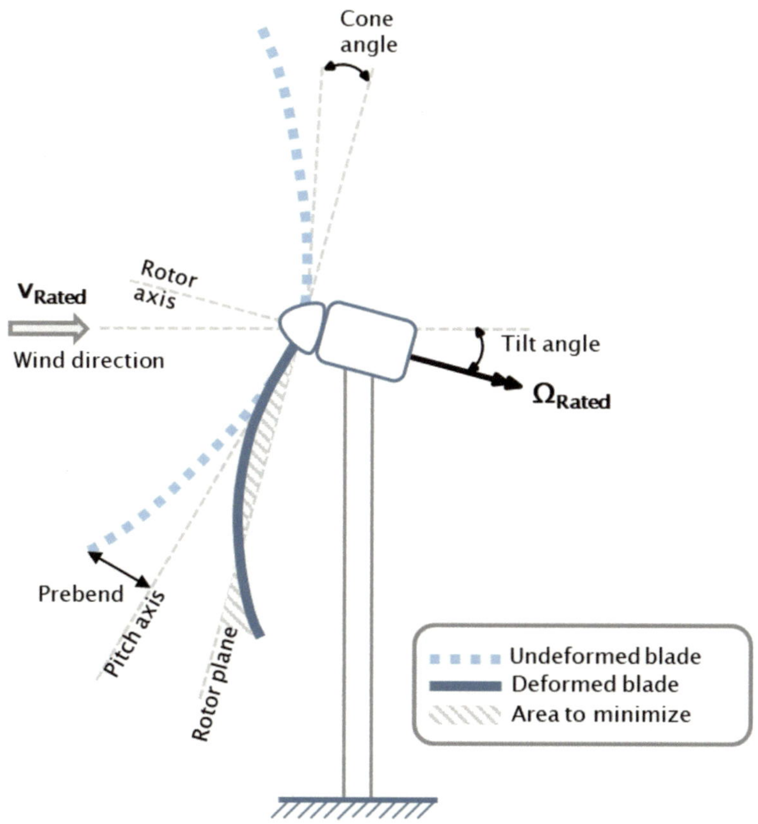

Figure 2.
Representation of the prebend design process.

With this in mind, the formalization of the prebend optimization subproblem can be stated according to Eqs. (13) and (14):

$$\left(\boldsymbol{p}_b^*, A_\delta^*\right) = \min_{\boldsymbol{p}_b}\left(A_\delta\left(\boldsymbol{p}_a^*, \boldsymbol{p}_b, \boldsymbol{p}_s, \boldsymbol{p}_g, \boldsymbol{r}_\epsilon^*\right)\right) \tag{13}$$

$$s.t.: \boldsymbol{g}_b\left(\boldsymbol{p}_b, \boldsymbol{p}_g\right) \leq \boldsymbol{0} \tag{14}$$

It must be noticed that, according to Eq. (14), the parameter A_δ is influenced not only by the shape of the prebend itself (defined by the parameters $\boldsymbol{p}_b$) but also by variables coming from the aerodynamic, structural, and global design. In particular, the PDS assumes that an aerodynamic design loop has been performed ahead, so that the optimal values of the aerodynamic design variables enter the module. Similarly, the PDS requires the knowledge of the regulation trajectory of the machine, computed at a previous step. With all this data available, the deflection of the blade is computed from simple static simulations in operating conditions, in order to design the prebend for a representative condition of the system without the need of time-consuming dynamic simulations. Eq. (15) lists a series of nonlinear constraints which are specific for the prebend design problem. These typically account for common manufacturing limitations such as maximum prebend at tip, maximum steepness of the prebend distribution, or application-specific requirements.

2.5 Structural design submodule

The structural design submodule (SDS) allows to perform the structural optimization of blades and tower with the aim of minimizing the initial capital costs (ICC) associated to the turbine manufacturing. This can be summarized in the optimization subproblem of Eqs. (15)–(17):

$$\left(\boldsymbol{p}_s^*, ICC^*\right) = \min_{\boldsymbol{p}_s}\left(ICC\left(\boldsymbol{p}_a^*, \boldsymbol{p}_b^*, \boldsymbol{p}_s, \boldsymbol{p}_g, \boldsymbol{r}_\epsilon^*\right)\right) \tag{15}$$

$$\boldsymbol{p}_s = \left[\boldsymbol{t}_b^{Fabrics}, \boldsymbol{t}_b^{Core}, \boldsymbol{t}_t, \boldsymbol{\rho}_t\right] \tag{16}$$

$$s.t.: \boldsymbol{g}_s\left(\boldsymbol{p}_a^*, \boldsymbol{p}_b^*, \boldsymbol{p}_s, \boldsymbol{p}_g, \boldsymbol{r}_\epsilon^*\right) \leq \boldsymbol{0} \tag{17}$$

According to Eq. (16), the computation of the ICC depends on the entire wind turbine model, so that all families of design variables enter the module. Since the SDS is the last submodule within the workflow of Cp-Max, it is assumed that the aerodynamic variables, as well as those related to prebend have been previously optimized. Eq. (17) shows the composition of the structural design variables: here, $\boldsymbol{t}_b^{Fabrics}$ is the thickness of all the blade structural elements, while $\boldsymbol{t}_b^{Core}$ accounts for the thickness of fillers (typically balsa or foam). The terms $\boldsymbol{t}_t$ and $\boldsymbol{\rho}_t$ account, respectively, for the internal thickness and radius of the tower segments. Starting from the internal layout and from the mechanical properties of the different materials, the SDS performs a sectional analysis based on the anisotropic beam theory proposed by Giavotto et al. [44]. This allows to compute, for each section, the fully populated mass and stiffness matrix. Those properties are then supplied to Cp-Lambda in order to update the structural properties of the blades and tower. Then, the ultimate loads and displacements are computed from an arbitrarily large set of design load cases (DLC) according to the international standards [45, 46], which typically include both deterministic and turbulent simulations. Once the envelope of ultimate and fatigue loads is known, a set of nonlinear constraints is enforced according to Eq. (18) in order to achieve a fully feasible structural solution. Classic constraints include resonance avoidance and frequency placement as well as ultimate stress/strain verification, a full fatigue analysis, a check on the maximum blade displacement, and a preliminary buckling analysis. The entire procedure is then repeated until the ICC is minimized. During optimization, it is also possible to perform detailed 3D FEM analyses in order to further refine the structural solution. The procedure is reported by Bottasso et al. [26].

2.6 Stability submodule

Stability analysis tools are of general utility for analyzing the causes of excessive vibration or for quantifying the effects of controllers on the closed-loop turbine behavior. Within the multidisciplinary design process of wind turbines, the stability analysis represents an important step to verify the proximity of flutter limits and drive the design onto regions in which all modes of interest are characterized by suitable damping levels.

The stability check can be performed on the entire turbine or on part of it, such as single blade. Within the design process, it is treated as a submodule external to the optimization loop. Hence, it is employed to verify the stability margins after the whole design or at the end of any MDL iteration.

Although stability is a well-known and studied problem for linear time invariant systems (LTI), some issue arises over the application of standard analysis methods to the case of wind turbines. In fact, as it often happens for rotating systems, wind turbines exhibit a behavior which is better comprehended within the theory of linear time periodic systems (LTP). This can be simply demonstrated by the fact that many physical phenomena entail wind turbine models characterized by periodic coefficients. Among all, one may mention the periodic stiffening induced by gravity, which compresses and extends the blades according to their position (upward or downward), or the asymmetry of the wind which entails periodically variable aerodynamic forces [47].

Periodic models can be rigorously studied through the Floquet theory [48], which shows that the stability of the system is uniquely defined by the characteristic multipliers, which are the eigenvalues of the so-called monodromy matrix (i.e., the transition matrix computed over one period). If the norm of each characteristic multiplier is lower than 1, then the system is asymptotically stable [49]. Since the computation of the monodromy matrix requires one integration of the equation of motion over a period for each state of the system, the application of the pure Floquet analysis to large and very large systems with thousands of degrees of freedom, as those characterizing modern high-fidelity multi-body models, is so much computationally expensive to result basically unaffordable even for current technology.

For this reason, many simplified methodologies, with much lower computational cost, often developed in the context of multi-body modeling of helicopters and wind turbines, have been proposed in literature with the aim of coping with Floquet analysis for very large systems. Among all, one can mention techniques which approximate the monodromy matrix with cascading Runge–Kutta solutions [50] or Chebyshev expansions [51] or algorithms which extract only the lowest characteristic multipliers from an incomplete transition matrix (noted as partial Floquet analysis [52]) or using Arnoldi's method (noted as implicit Floquet analysis [53]).

Dealing with wind turbine system stability, two other methods were recently proposed, one based on Coleman's transformation [54] and the other on a system identification procedure.

In the first method, Coleman's transformation is applied to a linearized state-space model of the turbine [3, 55, 56]. Such transformation has the effect of dramatically reducing the periodicity in the system matrices. The remaining periodic terms in the transformed system are then removed leading to an invariant system, which can be studied by a standard stability tool for LTI systems. Clearly, such an analysis is computationally affordable but presents two specific issues. Firstly, it cannot be applied to two-bladed wind turbines, as Coleman's transformation exists only for rotors with three or more blades. Secondarily, this approach is model-dependent, in the sense that any change of the topology of the turbine or any improvement in the nonlinear mathematical model has to be followed by an update of the linearization tool and of the implementation of Coleman's transformation itself. Due to the complexity of modern high-fidelity multi-body software tools, all these modifications may entail a significant effort.

Because of these issues and with the scope of having a MDO applicable to turbine models with any complexity, the approach based on system identification [47, 57, 58] has been chosen for the stability submodule. According to this approach, a simplified discrete periodic model of the periodic auto-regressive moving-average model with exogenous input (PARMAX) family is identified from input–output time histories of some suitable variables of interest. Then, the Floquet theory is applied to the reduced-order identified model, which accounts for only few states, i.e., those needed to capture the relevant behavior of the system. Such an approach, being based on input–output time histories, is independent from the simulator and from the model, in principle applicable to real turbines. Furthermore, since the Floquet analysis is performed on a system with few degrees of freedom, the entire approach results fully compliant with the periodic nature of the problem and affordable from a computational point of view.

From a practical standpoint, one has to generate input–output time histories from specific simulations where the modes of interest are mainly involved. Typically, one may introduce in the simulations some impulsive forces on the tower top and on the blade tip in order to excite the low-damped tower, whirling, and blade modes. Then the PARMAX model can be fitted on the measure of tower base and blade root bending moments, recorded after the end of the perturbation, where the aforementioned modes are well visible. With such an analysis, one may typically capture the most significant and low-damped modes of the wind turbine, which are the tower fore-aft and side-side modes, the forward and backward in-plane whirling, and the blade edgewise modes.

The damping ratios of those modes are then used within the MDO process to verify the stability margins and assess the possible proximity of the flutter boundaries. Using this approach, in [59] it was possible to prove that after a redesign of a blade including bend-twist-coupling, tower fore-aft and blade edgewise modes are characterized by a lower but still safe level of damping.

3. Applications

As discussed in the previous sections, the development of the design tool Cp-Max started in 2007, thanks to several funded industrial projects first in Italy and then also abroad. Over the years, up to four different blades have been designed and manufactured based on computations performed with this tool. Besides these industrial projects, other research activities have been carried out leading to the implementation of the Cp-Max code, as discussed in the previous sections. In the next paragraphs, we present some results aiming to highlight some features of the tool.

3.1 Design of a 20 MW wind turbine

Within the European FP7 project INNWIND.EU [60], significant efforts have been dedicated to the development of technological solutions able to reduce the COE. Among these, the synergy between active and passive load alleviation systems represents certainly an interesting solution for the design of lightweight rotors. These solutions have demonstrated how to smartly increase the size and power of future wind turbines. For this reason, a conceptual design of a 20 MW wind turbine has represented an important step, and the design tool Cp-Max has been demonstrated to have all the features needed to obtain a good preliminary design without upscaling laws.

From a historical perspective, as Sartori shows in details [61], the rotor design started from a reference wind turbine obtained from the INNWIND.EU 10 MW. At first, classical scaling laws have been applied to upscale the reference 10 MW turbine to 20 MW. Afterward, Cp-Max has been employed to produce a realistic structure for the upscaled model. These two steps provided a baseline model (in the following referred to as "baseline 20 MW"), which was optimized afterward. In order to deeply understand the effect of each submodule presented in Section 2, we adopted a parametric approach where we performed three design steps: (I) a prebend analysis, (II) a bend-twist coupling analysis, and (III) a solidity analysis. At the end of each step, the optimal configuration was taken as the starting point of the following one. It must be stressed that a full structural redesign of each solution under investigation has been conducted, so that all the configurations analyzed satisfy the same design constraints and hence may be fairly compared.

First, the prebend submodule of Section 2.4,was used to define the prebend distribution. In a sensitivity analysis, the main parameter was the tip value of prebend distribution, which was assumed equal to 2 and 4 m. The final spanwise distribution computed by the Cp-Max submodule is shown in **Figure 3**. The latter solution has been selected for the following step due to its (slightly) better capability of reducing the COE, as can be noticed in **Figure 4**. This COE reduction is due to the mass drop caused by the higher blade-to-tower distance, which allows the blade itself to be more flexible.

The second step was the introduction of a passive load alleviation system in the blade. This has been achieved with a coupling between the bending and the torsion deformation of the blade, the so-called bend-twist coupling (BTC), as previously presented in [62] for a smaller blade. This coupling is here obtained by moving the fiber direction of the unidirectional material in the spar caps away from the pitch axis.

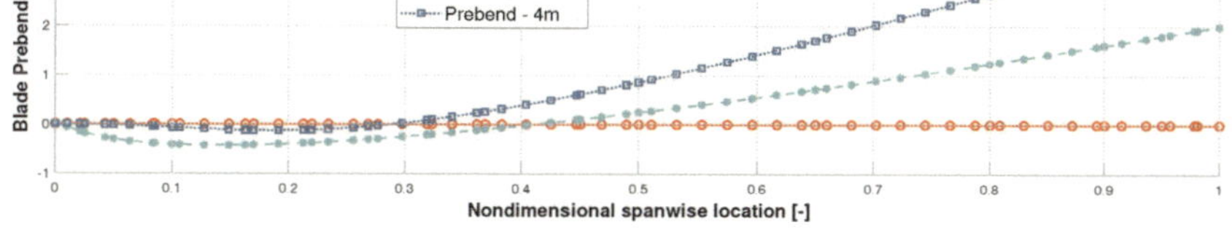

Figure 3.
Spanwise optimal prebend distribution.

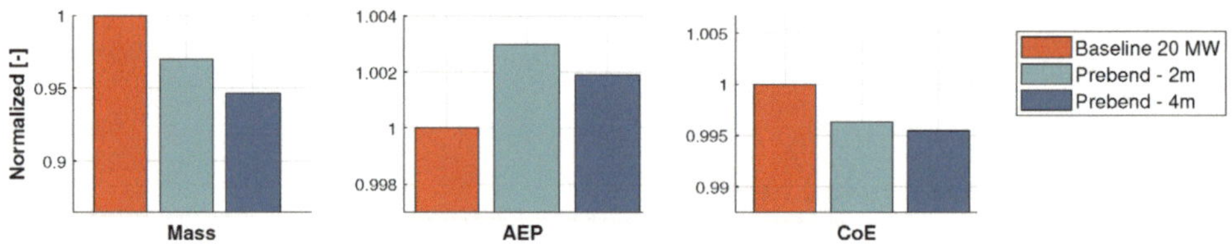

Figure 4.
Nondimensional mass, AEP, and COE of the optimized prebended blades.

Again, a sensitivity analysis has been conducted to deeply understand the effects of the BTC angle, assumed as the parameter of the analysis, on the wind turbine loads and on global parameters as well, i.e., the blade mass, the AEP, and the COE. The values considered here, selected from previous analyses performed on other blades, are 4, 6, and 8°. The effects of the BTC are mainly felt on the fatigue loads of wind turbine subcomponents (as shown in [61]) and on the actuator duty cycle, so that without any further action (i.e., stretching of the blade, redesign of the tower/hub systems, etc.), the COE is almost unchanged, as one can see in **Figure 5**. However, the same figure shows that the blade mass may start to rise when over-increasing the fiber angle. This is because the higher the fiber angle, the higher the spar cap thickness needed to restore the flapwise stiffness, which is constrained in the design process by the maximum blade deflection and by the blade frequency placement with respect to the rotor speed. Finally, the AEP picture shows that the BTC tends to reduce the power output of the rotor. This is due to the increased torsional deformation of the blade which, during the normal operation of the wind turbine, changes the optimal local angle of attack. This effect may be bypassed by a fine tuning of the control law as shown in Section 2.3.

The last step involves an analysis of the effect of solidity, starting from the optimal 6° BTC blade, which represents the best compromise between the load and mass reductions. This solidity analysis required the aerodynamic submodule (2.2) to define the external geometry, together with the structural submodule (2.5) to size the internal geometry for each chord distribution. The solidity has been reduced to 98, 96, and 94%, and the effects on the fatigue loads of these reductions are shown in **Figure 6**, whereas the effects on the global parameters are featured in **Figure 7**.

As one can see, the reduction in the blade solidity generates lower fatigue loads in the nonrotating subsystems of the wind turbine (i.e., nacelle and tower), while it increases the loads at the root of the blades. This is because the reduction of the chord, and hence in the blade thickness, must be compensated by an increase of the spar cap thickness to restore the flapwise stiffness, for fixed airfoils characteristics. This increase in the total blade mass (see **Figure 7**) in turn increases blade loads. The overall effects on the AEP and COE are, on the other side, negligible. Therefore, the 96% solution is here assumed as the best compromise considering all performance indicators.

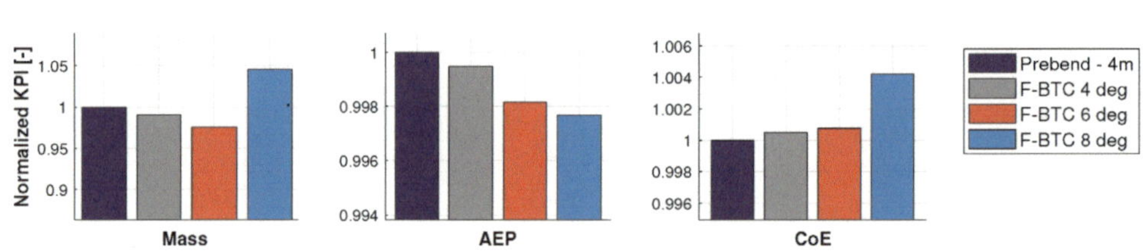

Figure 5.
Nondimensional mass, AEP, and COE of the optimized bend-twist coupled blades.

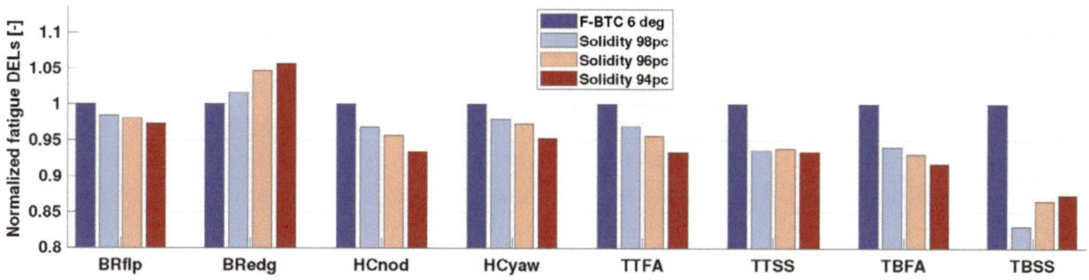

Figure 6.
Nondimensional fatigue loads (DELs) comparison. BR = blade root, TB = tower base, TT = tower top, and HC=hub center.

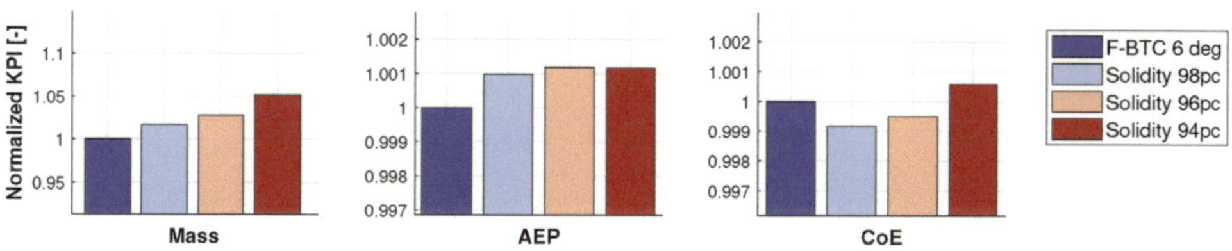

Figure 7.
Mass, AEP, and COE of the optimized blades.

	Units	Baseline	Optimal	Change
		20 MW	20 MW	
Total blade mass	[ton]	113.5	107.8	−5.05%
AEP	[GWh]	91.6	91.7	+0.12%
COE	[EUR/MWh]	84.9	84.6	−0.42%
Blade root flap DEL	[MNm]	83.8	75.6	−9.79%
Hub nodding DEL	[MNm]	53.6	46.7	−12.83%
Tower base FA DEL	[MNm]	278.5	271.6	−2.48%
Tower base SS DEL	[MNm]	204.4	164.9	−19.2%

Table 1.
Performance comparison between the baseline and the optimal 20 MW rotors.

Finally, an overall comparison between the baseline rotor and the optimized one, equipped with a 6° BTC, a 4-m-tip prebend, and a reduced 98% solidity, is shown in **Table 1**. The comparison highlights only a slight decrease in the COE (of about 0.42%), but it is important to stress here that this reduction is also coupled with lower tower and hub loads. This trend suggests the chance to redesign these subcomponents and/or to design a longer blade able to transmit to the fixed subsystems of the turbine the same baseline loads. Both of these solutions, or a combination of them, may hence further decrease the COE.

3.2 Wind turbine under wind farm control

In recent years, many efforts have been placed in research activities related to the development of wind farm controllers. The idea behind this topic is to look for optimal wind farm operations in order to maximize the power output of the farm which, in turn, is strictly connected to the minimization of the wake interfaces between the machines. In the analyses related to the development of these wind farm control algorithms, much attention has been devoted to the verification of the AEP and the fatigue loads, but little has been done about the analysis of the impact of these controllers on the load envelope of the single machine and therefore on its design. In fact, the wind farm controller techniques may alter the normal functionality of the single wind turbine and hence change the critical design conditions. This may cause design loads and blade deflections to exceed the design constraints.

In this paragraph we present, as a further example of a possible use of the multidisciplinary design tool Cp-Max, the redesign of the 10 MW INNWIND.EU wind turbine, in case it is subjected to a wind farm control technique, named active wake mixing (AWM). This is a control scenario where the blades are pitched to create a dynamic rotor thrust, with the effect of sucking undisturbed air into the wake, so as to reenergize it. In this research, the AWM is obtained by periodically changing the collective pitch angle [63] as

$$\beta_{\mathrm{AWM}} = A_{\mathrm{AWM}} \sin\left(2\pi f_{\mathrm{AWM}} t\right), \tag{18}$$

where β_{AWM} is the pitch setting imposed by wind farm control, which is superimposed to the pitch of the trimmer, A_{AWM} is the related amplitude, and f_{AWM} is the frequency. Typically, rather than in terms of frequency f, the effect of AWM is viewed in terms of the dimensionless Strouhal number S_t, defined as

$$S_t = \frac{f_{\mathrm{AWM}} D}{U_\infty}, \tag{19}$$

being U_∞ the undisturbed wind velocity and D the rotor diameter. Parametric aeroelastic analyses conducted on this wind turbine showed that the maximum loads on the blade, and hence the maximum deflection, may exceed the design values especially for higher amplitude of pitch actuation A_{AWM} and Strouhal numbers. This means that the rotor may need to be redesigned when the wind farm controller is applied to the farm. The design problem here requires to include in the control system synthesis of Section 2.3 the AWM activity in order to include in all DLCs the effect of the wind farm controllers. Energizing the wake is useful only at low wind speed, because in the above-rated region, the wake still maintains enough energy to allow the downstream wind turbine to operate at full power. For this reason, the AWM in this research is switched off in all the simulations where the mean hub wind speed is higher than 15 m/s. Clearly, the wind farm control is not included in the DLCs where the wind turbine is parked. Nevertheless, the latter simulations must be included in the design process since they may turn out to be design-driving loads. When this happens, i.e., when the design-driving loads arise from not-controlled cases, the wind farm controllers do not affect the wind turbine rotor. The two parameters of the AWM, after a sensitivity analysis, have been defined as $A_{\mathrm{AWM}} = 2°$ and $S_t = 0.5$, which corresponds to a good compromise between the need to energize the wake and to avoid excessive loads on the upwind turbine. To better understand the effects of the wind farm controller on the rotor design, we first redesign the INNWIND.EU 10 MW wind turbine without any farm controllers. In so doing, we are sure the new 10 MW baseline will be the result of a design process considering the same trimmer (2.3), the same DLCs, the same aeroelastic model, and the same design approach and constraints. Starting from this new 10 MW baseline, we restart the Cp-Max design loop including in the process also the DLCs with the AWM controller. It is important to stress here that also the simulations without the farm controller must be considered, since they may generate higher loads and, in general, the AWM may be switched off for selected wind

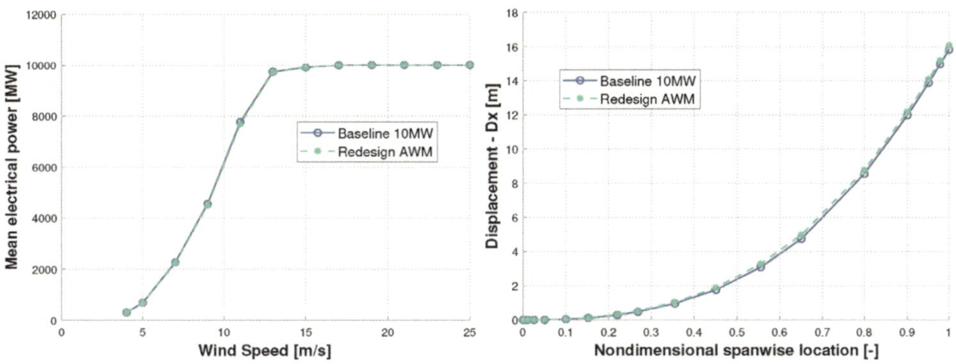

Figure 8.
Power curves (left) and maximum blade deflections (right) for the baseline model and the optimized one.

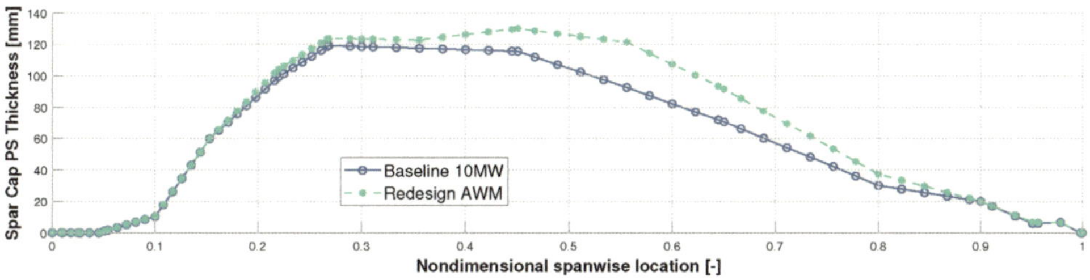

Figure 9.
Spanwise pressure-side spar cap thickness distributions.

directions. In **Figure 8**, the power curves' comparison between the baseline wind turbine and the optimized one is displayed. As one can see, in the partial-power region, despite the AWM actuation, which causes the blade to move out of its optimal pitch value, the power curve remains close to the baseline. **Figure 8** shows the maximum blade deflections to be very close to each other. This result comes directly from the optimization process, where the blade tip deflection is for this rotor an active constraint. However, these maximum deflections are obtained with a different blade stiffness. Due to the higher loads under the wind farm controllers, Cp-Max has to increase the spar cap thicknesses in the structural submodule (2.5) as

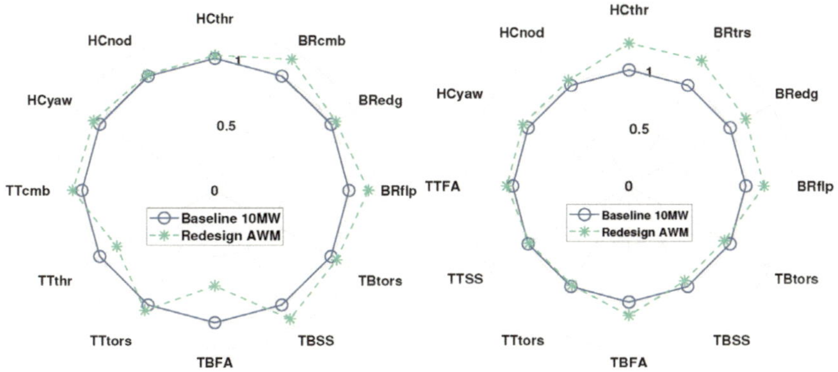

Figure 10.
Fatigue DEL (left) and ultimate loads (right) comparison of the redesign AWM against the baseline (the latter normalized to one). BR = blade root, TB = tower base, TT = tower top and HC = hub center.

	Baseline	AWM-redesigned	Variation
Blade mass	40,643 kg	45,436 kg	+11.8%
AEP	45.86 GWh	45.63 GWh	−0.5%
COE	89.42 EUR/MWh	90.22 EUR/MWh	+0.89%

Table 2.
Baseline rotor vs. AWM-redesigned rotor.

shown in **Figure 9**. Finally, **Figure 10** shows the comparison between the normalized loads on the baseline and optimized rotor. In this figure, on the left plot, one can see the fatigue loads (DELs) and on the right the maximum ones. BR refers to blade root, TB to tower base, TT to tower top, and HC to hub center. The main conclusion from this picture is that the loads to which the single wind turbine is subjected when operating under AWM control may increase, and this generates a heavier structure as shown in **Table 2**.

4. Conclusions

In this chapter, we proposed the use of a multidisciplinary design optimization framework to design next-generation wind turbines. We showed how such algorithms require to compromise between different aspects pertaining to the optimization, such as scope, number of variables, merit functions, and modeling fidelity. In this context, we presented a multilevel algorithm which is able to perform the complete design of a wind turbine, through the interface of several optimization modules.

Based on the results presented in this chapter and on an extensive practice on MDO, not shown here for the sake of brevity, the following conclusions can be drawn:

- A wind turbine design conducted in a multidisciplinary environment is necessary to fully capture all the couplings and interconnections existing in the nonlinear response of the turbine. In fact, a rotor which is optimal only from an aerodynamics standpoint may be far from optimal when it comes to the overall design (i.e., structure, control, cost of energy).

- The proposed modular architecture offers a sufficient flexibility to treat a wide number of design variables without sacrificing the accuracy of the physical description. This allows to derive driving loads and deflections directly from fully resolved design load cases, so that the optimal solution is automatically standard-compliant.

- The use of a model-based control eases the optimization procedure since the control parameters are automatically adjusted to the evolving model within the design process.

- As the entire proposed loop operates on a full multi-body model of the wind turbine, it is immediately possible to perform specific further analyses, thus improving the confidence of the solution. For example, the assessment of the stability of the system can be run to ensure that the optimal solution does not show resonance problems.

- The procedure hitherto exposed can be employed for a variety of applications, which include preliminary wind turbine or detailed component design, trade-off analyses, verification of the impact of selected control laws on cost of energy, and design of unconventional configurations.

Conflict of interest

The authors declare that there is no conflict of interest.

Author details

Luca Sartori, Stefano Cacciola, Alessandro Croce*
and Carlo Emanuele Dionigi Riboldi
Department of Aerospace Science and Technology, Politecnico di Milano, Milano, Italy

*Address all correspondence to: alessandro.croce@polimi.it

All authors are contributed equally.

References

[1] International Renewable Energy Agency (IRENA). Renewable Power Generation Costs in 2017. Technical Report. Abu Dhabi, UAE: IRENA; 2018

[2] Stettner M, Reijerkerk MJ, Lünenschloß A, Riziotis V, Croce A, Sartori L, et al. Stall-induced vibrations of the AVATAR rotor blade. Journal of Physics: Conference Series. 2016;**753**(4): 042019

[3] Hansen MH. Aeroelastic stability analysis of wind turbines using an eigenvalue approach. Wind Energy. 2004;7(2):133-143

[4] Buck JA, Garvey SD. Redefining the design objectives of large offshore wind turbine rotors. Wind Energy;**18**(5): 835-850

[5] Ning AS, Damiani R, Moriarty PJ. Objectives and constraints for wind turbine optimization. Journal of Solar Energy Engineering. 2014;**136** (4):041010

[6] Bortolotti P, Sartori L, Croce A, Bottasso CL. Multi-MW wind turbine CoE reduction via a multi-disciplinary design process. In: Proceedings of the EWEA 2015; 2015

[7] Chaviaropoulos PK, Beurskens HJM, Voutsinas SG. Moving towards large(r) rotors is that a good idea? In: Proceedings of EWEA 2013; Vienna, Austria; 2013

[8] Kenway G, Martins JRRA. Aerostructural shape optimization of wind turbine blades considering site-specific winds. In: 12th AIAA/ISSMO Multidisciplinary Analysis and Optimization Conference; Victoria, British Columbia, Canada; 2008

[9] Xudong W, Shen WZ, Zhu WJ, Sørensen JN, Jin C. Shape optimization of wind turbine blades. Wind Energy. 2009;**12**(8):781-803

[10] Pourrajabian A, Nazmi Afshar PA, Ahmadizadeh M, Wood D. Aero-structural design and optimization of a small wind turbine blade. Renewable Energy. 2016;**87**:837-848

[11] Bottasso CL, Campagnolo F, Croce A. Multi-disciplinary constrained optimization of wind turbines. Multibody System Dynamics. 2012;**27**:21-53

[12] Sessarego M, Shen WZ. Aero-structural optimization of wind turbine blades using a reduced set of design load cases including turbulence. Journal of Physics: Conference Series. 2018;**1037**(4): 042018

[13] Pavese C, Tibaldi C, Larsen TJ, Kim T, Thomsen K. Reduced design load basis for ultimate blade loads estimation in multidisciplinary design optimization frameworks. Journal of Physics: Conference Series. 2016;**753**(6):062005

[14] Zhu J, Cai X, Gu R. Aerodynamic and structural integrated optimization design of horizontal-axis wind turbine blades. Energies. 2016;**9**(2):66

[15] Dal Monte A, De Betta S, Raciti Castelli M, Benini E. Proposal for a coupled aerodynamic structural wind turbine blade optimization. Composite Structures. 2017;**159**:144-156

[16] Yang H, Chen J, Pang X, Chen G. A new aero-structural optimization method for wind turbine blades used in low wind speed areas. Composite Structures. 2019;**207**:446-459

[17] Richards PW, Griffith DT, Hodges DH. Aeroelastic design of large wind turbine blades considering damage tolerance. Wind Energy. 2016;**20**(1): 159-170

[18] Pavese C, Tibaldi C, Zahle F, Kim T. Aeroelastic multidisciplinary design optimization of a swept wind turbine

blade. Wind Energy. 2017;**20**(12): 1941-1953

[19] Fuglsang P, Bak C, Schepers JG, Bulder B, Cockerill TT, Claiden P, et al. Site-specific design optimization of wind turbines. Wind Energy. 2002;**5**(4): 261-279

[20] Maki K, Sbragio R, Vlahopoulos N. System design of a wind turbine using a multi-level optimization approach. Renewable Energy. 2012;**43**:101-110

[21] Ashuri T, Zaaijer MB, Martins JRRA, van Bussel GJW, van Kuik GAM. Multidisciplinary design optimization of offshore wind turbines for minimum levelized cost of energy. Renewable Energy. 2014;**68**(Supplement C):893-905

[22] Ashuri T, Zaaijer MB, Martins JRRA, Zhang J. Multidisciplinary design optimization of large wind turbines—Technical, economic, and design challenges. Energy Conversion and Management. 2016;**123**:56-70

[23] Dykes K, Ning A, King R, Graf P, Scott G, Veers P. Sensitivity analysis of wind plant performance to key turbine design parameters: A systems engineering approach. In: AIAA SciTech 2014; National Harbor, Maryland; 2014

[24] Gray J, Moore KT, Naylor BA. OpenMDAO: An open source framework for multidisciplinary analysis and optimization. In: 13th AIAA/ISSMO Multidisciplinary Analysis Optimization Conference; 2010

[25] Zahle F, Tibaldi C, Verelst DR, Bak C, Bitsche R, Blasques JPAA. Aero-elastic optimization of a 10 MW wind turbine. American Institute of Aeronautics & Astronautics. 2015;**1**: 201-223

[26] Bottasso CL, Campagnolo F, Croce A, Dilli S, Gualdoni F, Nielsen MB. Structural optimization of wind turbine rotor blades by multilevel sectional/multibody/3D-FEM analysis. Multibody System Dynamics. 2014;**32**: 87-116

[27] Gualdoni F. Design optimization of wind turbines [PhD thesis]. Milano, Italy: Politecnico di Milano; 2014

[28] Bortolotti P, Bottasso CL, Croce A. Combined preliminary-detailed design of wind turbines. Wind Energy Science. 2016;**1**(1):71-88

[29] Sartori L. System design of lightweight wind turbine rotors [PhD thesis]. Milano, Italy: Politecnico di Milano; 2019

[30] Bottasso CL, Croce A. Cp-Lambda. A Code for Performance, Loads, Aeroelasticity by Multi-Body Dynamics Analysis. Technical Report. Milano, Italy: Politecnico di Milano; 2010–2017

[31] Fingersh L, Hand M, Laxson A. Wind Turbine Design Cost and Scaling Model. Technical Report: NREL/TP-500-40566. Boulder, CO: National Renewable Energy Laboratory; 2006

[32] Bottasso CL, Croce A, Riboldi CED, Nam Y. Multi-layer control architecture for the reduction of deterministic and non-deterministic loads on wind turbines. Renewable Energy. 2013;**51**: 159-169

[33] Bossanyi E. Wind turbine control for load reduction. Wind Energy. 2003; **6**(3):229-244

[34] Bottasso C, Croce A, Savini B, Sirchi W, Trainelli L. Aero-servo-elastic modeling and control of wind turbines using finite element multibody procedures. In: Proceedings of the ECCOMAS Multibody Dynamics 2005 Thematic Conference, Vol. 16; Madrid, Spain; June 21–24, 2006. pp. 291-308

[35] Riboldi CED. Advanced control laws for variable-speed wind turbines and

supporting enabling technologies [PhD thesis]. Milano, Italy: Politecnico di Milano; 2012

[36] Bottasso CL, Pizzinelli P, Riboldi CED, Tasca L. LiDAR-enabled model predictive control of wind turbines with real-time capabilities. Renewable Energy. 2014;**71**: 442-452

[37] Bottasso CL, Riboldi CED. Estimation of wind misalignment and vertical shear from blade loads. Renewable Energy. 2014;**62**:293-302

[38] Bottasso CL, Riboldi CED. Validation of a wind misalignment observer using field test data. Renewable Energy. 2015;**74**:298-306

[39] Bertelè M, Bottasso CL, Cacciola S, Daher Adegas F, Delport S. Wind inflow observation from load harmonics. Wind Energy Science. 2017;**2**(2):615-640

[40] Cacciola S, Riboldi CED, Croce A. A new decentralized pitch control scheme for wind turbines. IFAC-Papers Online. 2017;**50**:9908-9913

[41] Riboldi CED, Cacciola S. Individual pitch control for 2-bladed wind turbines via multiblade multilag transformation. Wind Energy. 2017;**20**(12):1955-1969

[42] Cacciola S, Riboldi CED. Equalizing aerodynamic blade loads through individual pitch control via multiblade multilag transformation. Journal of Solar Energy Engineering, Transactions of the ASME. 2017;**139**(6):061008

[43] Riboldi CED. On the optimal tuning of individual pitch control for horizontal-axis wind turbines. Wind Engineering. 2016;**40**(4):398-416

[44] Giavotto V, Borri M, Mantegazza P, Ghiringhelli G, Carmaschi V, Maffioli GC, et al. Anisotropic beam theory and applications. Computers & Structures. 1983;**16**(1):403-413

[45] International Electrotechnical Commission. IEC 61400-1 Wind Turbines. Part 1: Design Requirements. 3rd ed. Geneva, Switzerland: International Electrotechnical Commission; 2006

[46] Germanischer Lloyd Industrial Services GmbH. Guideline for the Certification of Wind Turbines. 2010

[47] Bottasso CL, Cacciola S. Model-independent periodic stability analysis of wind turbines. Wind Energy. 2015; **18**(5):865-887

[48] Floquet G. Sur les équations différentielles linéaires á coefficients périodiques. Annales Scientifiques de l'E.N.S. 1883;**2**:47-88

[49] Bittanti S, Colaneri P. Periodic Systems Filtering and Control. London: SpringerVerlag; 2009

[50] Friedmann P, Hammond CE, Woo T-H. Efficient numerical treatment of periodic systems with application to stability problems. International Journal for Numerical Methods in Engineering. 1977;**11**(7):1117-1136

[51] Sinha SC, Pandiyan R. Analysis of quasilinear dynamical systems with periodic coefficients via Liapunov-Floquet transformation. International Journal of Non-Linear Mechanics. 1994; **29**(5):687-702

[52] Bauchau OA, Wang J. Efficient and robust approaches to the stability analysis of large multibody systems. Journal of Computational and Nonlinear Dynamics. 2007;**3**(1):011001

[53] Bauchau OA, Nikishkov YG. An implicit Floquet analysis for rotorcraft stability evaluation. Journal of the American Helicopter Society. 2001; **46**(3):200-209

[54] Coleman R, Feingold A. Theory of Self-Excited Mechanical Oscillations of

Helicopter Rotors with Hinged Blades. Technical Report: NACA Technical Report 1351. Langley Field, VA, United States: National Advisory Committee for Aeronautics; 1958

[55] Bir G. Multi-blade coordinate transformation and its application to wind turbine analysis. In: 46th AIAA Aerospace Sciences Meeting and Exhibit; Reno, NV; January 7–10, 2008

[56] Skjoldan P, Hansen M. On the similarity of the Coleman and Lyapunov-Floquet transformation for modal analysis of blade rotor structures. Journal of Sound and Vibration. 2009; **327**(3):424-439

[57] Bottasso CL, Cacciola S, Riva R. Floquet stability analysis of wind turbines using input-output models. In: 32nd ASME Wind Energy Symposium; National Harbor, MD; January 13–17, 2014

[58] Riva R, Cacciola S, Bottasso CL. Periodic stability analysis of wind turbines operating in turbulent wind conditions. Wind Energy Science. 2016; **1**(2):177-203

[59] Riva R, Spinelli M, Sartori L, Cacciola S, Croce A. Stability analysis of wind turbines with bend-twist coupled blades. Journal of Physics: Conference Series. 2018;**1037**:062014

[60] Jensen PH, Natarajan A. INNWIND.EU. Overview of project and recent results, Technical Report, DTU, Copenhagen, 2014. www.innwind.eu

[61] Sartori L, Bellini F, Croce A, Bottasso CL. Preliminary design and optimization of a 20MW reference wind turbine. Journal of Physics: Conference Series. 2018;**1037**:042003

[62] Bottasso CL, Campagnolo F, Croce A, Tibaldi C. Optimization-based study of bend-twist coupled rotor blades for passive and integrated passive/active load alleviation. Wind Energy. 2013; **16**(8):1149-1166

[63] Frederik J, Weber R, Cacciola S, Campagnolo F, Croce A, Bottasso CL, et al. Periodic dynamic induction control of wind farms: Proving the potential in simulations and wind tunnel experiments. Wind Energy Science Discussions. DOI: 10.5194/wes-2019-50. 2019;**2019**:1-18

Chapter 3

Wind Turbine Airfoil Boundary Layer Optimization Using Genetic Algorithm with 3D Rotational Augmentation

Youjin Kim, Galih Bangga and Antonio Delgado

Abstract

The airfoil shape of horizontal axis wind turbine (HAWT) blade is optimized using genetic algorithm (GA). The algorithm is set to find the final airfoil shape with the highest gliding ratio (GR) and larger laminar boundary layer regime along the airfoil surface. The main aim is to find the best airfoil shape of higher lift coefficient with reduced drag in boundary layer from the reference airfoil shape. A 3D correction law is applied to model the effect of optimized airfoil in 3D rotational augmented situation. The thrust and power curves are generated by the blade element (BEM) and free vortex (FV) codes with 3D and loss correction. The higher power production is given when the wind turbine blades are designed using the optimized airfoil. This increment is thought to be made from the efficiency caused by the reduced separation bubbles from reduced turbulent boundary layer and 3D rotational augmentation. To validate its effectiveness in case of soiled condition, the aerodynamic parameters of airfoils are recalculated by enforcing the airfoil to undergo earlier transition, which models the leading edge roughness. The results indicate the soiled condition that does not affect the aerodynamic efficiency of the airfoil due to the positive effect of 3D rotation augmentation.

Keywords: optimization, genetic algorithm, airfoil, wind turbine blade, aerodynamics, rotational augmentation

1. Introduction

The purpose of optimization is to find optimal solutions in scientific or engineering problems. The optimization can also be applied to many wind turbine problems. According to different objectives, constraints, algorithms, tools, and models, various types of optimization solutions are possible. In literature, Jabaraj and Iniyan [1] mentioned optimization models in the wind conversion system with different modelings such as planning model, energy supply-demand model, and forecasting model in 2000. The computational optimization algorithm is mentioned by Baños [2] for solving problems in renewable energy. The artificial intelligence methods are used for forecasting wind speed and power by Lei [3]. The updated numerical simulations and other technology trends are reported in 2013 by Miller [4]. The wind farm layout optimization is reviewed by Serrano González [5]. Mostly objective functions are the minimization of cost of energy, maximization of the power production, minimization of the blade mass, and so on [6–20]. The geometrical, aerodynamic, and physical aspects are to be researched when the constraints are in the field of turbine design. Especially if the blade section shape, called airfoil, is targeted for aerodynamic optimization, stochastic optimization is preferred than the gradient one due to the usefulness in shape randomness [21]. Among many stochastic methods, Evolutionary Computation (EC) is considered in this paper.

The biological concept that survives the fittest individual in the environment among the others is applied in EC [22]. GA, which is one of ECs, runs until it finds the fitness individual with the highest fittest level from the given objective function. It considers the individual solution candidate as the gene, which is the concept of reproduction unit of Mendel [23], and the individuals are exposed to different strategies of genetics to make another generation of the solution candidate pool. The reproduction strategy includes reproduction, crossover, and mutation [24]. The airfoil shapes within given upper and lower bounds make possible solution candidates. The algorithm runs for the fittest individual in objective function f(x) = {GR + Xtr}, which means the algorithm finds the airfoil with the highest Gliding Ratio (GR) and the latest transition point (Xtr), in other words, larger laminar boundary layer regime on the airfoil surface. It finds the best airfoil of the highest lift coefficient (Cl) and lowest drag coefficient (Cd) in the generated airfoil candidate pool [25].

Two airfoils are compared to show their GR, Cl, Cd, and Power production in the wind turbine unit. As the algorithm is run with calculations of the airfoil in 2D, the correction law to consider the 3D effects and Rfoil software is applied [26, 27]. The 3D rotational effect of rotating machines has been found by many. The lift coefficient of the fan blade was found to be three times higher by Himmelskamp in [28]. The lift coefficient of a wind turbine blade was also found to be higher at the inboard sections of the blade by the experiments of Ronsten and Bruining [29, 30]. Later, correction laws for the 3D effect were tried by numerical investigations. The quasi-3D approaches by Hansen [31] and Snel [32, 33] led to the quasi-3D Navier-Stokes mode [34], which confirmed its validity by Shen and Soerensen [35].

As the rotation of the rotor was found to reduce separation and transition by the Coriolis force [35], the 3D correction terms are considered together with the optimized airfoil shape. By comparing the results corrected by the 3D correction law, the effect of optimized shape for higher GR and larger laminar boundary layer in lift coefficient and power production under 3D rotational effects can be deduced. The rotor Power and thrust curves show the combined effect of optimized airfoil on lifting efficiency in the blade unit by BEM theory [36] and FV method [37]. The thrust and power curve comparison leads us to see the effect of the blade lift efficiency increment caused by the optimized airfoil. Moreover, as the total power from the rotor is considered based on each section of blade annulus in BEM, the effects each 3D corrected aerodynamic parameter values of blade section with optimized airfoils are combined to contribute to the increase of power production. Moreover, lifting line of FV method, which calculates total external force and the lift of vortices strength, is also based on the lifting lines on the divided blade segments [37]. Its power calculation also reveals the gathered influence of increased efficiency of the optimized airfoil in sections of the blade.

The compensation for some missing correction laws in power calculations from BEM and FV codes with 3D correction law is possible with the code B-Go. The code B-Go is validated with experimental and computational results, which confirm its reliability based on several correction terms, including tip loss correction, and flow conditions such as massive flow separation takes place [38].

Another realistic rotational situation of airfoil is soiled condition [39]. The leading edge of the airfoil is exposed to debris, dirt, soil, and pollution, and so on in the real situation. The Cl parameter of airfoils is calculated with forced transition. As this contamination is known to decrease the rated and maximum power [40], the roughness sensitivity of airfoil is of importance in the generation of new airfoil. As the contamination usually forces the transition of boundary layer to the leading edge of airfoil, this study made optimized and reference airfoils to have Xtr = 0.05 on the suction side and Xtr = 0.1 on the pressure side as it is recommended in the work of [41].

This chapter illustrates the results of the airfoil and design shape of turbine blade in Section 2, followed by the aerodynamic characteristics in Section 3. The power calculation of the turbine blade with optimized airfoil with corrected BEM and B-Go is shown in Section 4. The airfoil validity in soil condition is elaborated in Section 5. The summarization of the results and their interpretation are shown in Section 6.

2. Airfoil and blade design

The optimized airfoil called S809gx is generated with the settled GA algorithm at the Reynolds number (Re) 10^6 [25]. The reference airfoil and wind turbine are NREL Phase VI [42, 43]. The difference of thickness, maximum thickness, and maximum camber values are negligible, as shown in **Table 1** and **Figure 1**.

Because the optimized airfoil is found from the algorithm run to finish at the higher GR and larger Xtr point value at specific angle of attack (AOA) [25], airfoil S809gx shows to have 121% higher GR value, 168% larger laminar boundary layer region on the suction side of airfoil, and 125% larger laminar boundary layer on

	S809	S809gx
Thickness (%)	20.99	20.3
Max. thickness possible (%)	38.3	38.7
Max. camber (%)	0.99	0.87
Max. camber possible (%)	83.3	43.6

Table 1.
Airfoil properties [25].

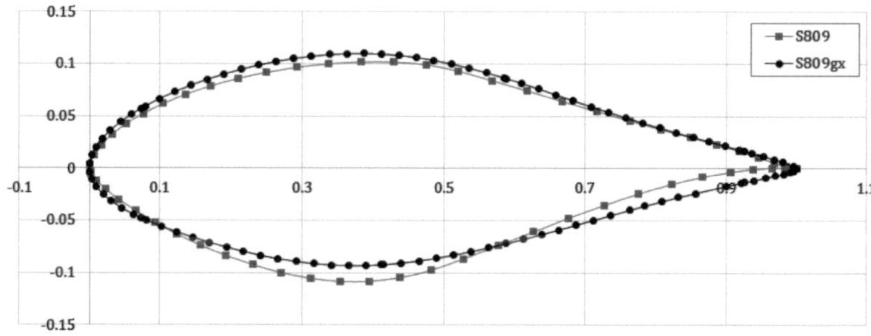

Figure 1.
Optimized airfoil S809gx and reference S809 [25].

the pressure side at AOA 7°, as depicted in **Table 2**. It also indicates 140% higher GR values, 400% larger laminar boundary layer region on suction side, and 162% larger laminar boundary layer on the pressure side at AOA 21.5°. Those AOA values are chosen as the representative angle for fully attached and stall separation flow around airfoil.

The shape factor H of boundary layer [44] is plotted for both airfoils at targeted angle of attack. The optimized airfoil shows H values to be lower than 2.0 at $x/c > 0.3$

	S809gx	**S809**	**S809gx**	**S809**
AOA (°)	7	7	21.5	21.5
Cl	0.899	0.8793	1.0264	0.9149
Cd	0.011	0.0127	0.1566	0.1958
GR	85.29	69.50	6.553	4.672
Xtr (suction side)	0.272	0.162	0.016	0.004
Xtr (pressure side)	0.677	0.540	1.000	0.616

Table 2.
GR and Xtr values of airfoils (Rfoil).

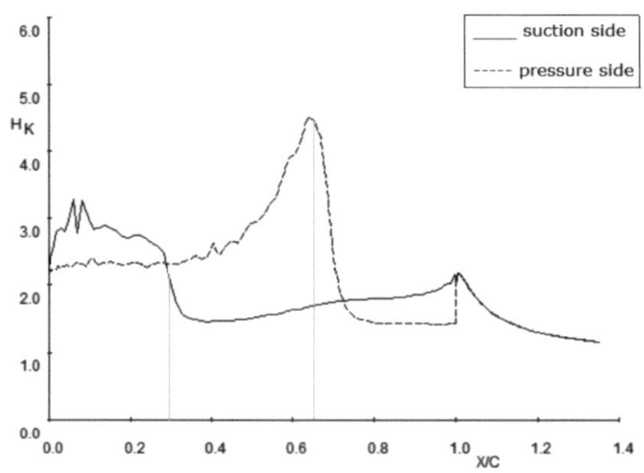

Figure 2.
Shape factor of boundary layer of S809gx at AOA = 7°.

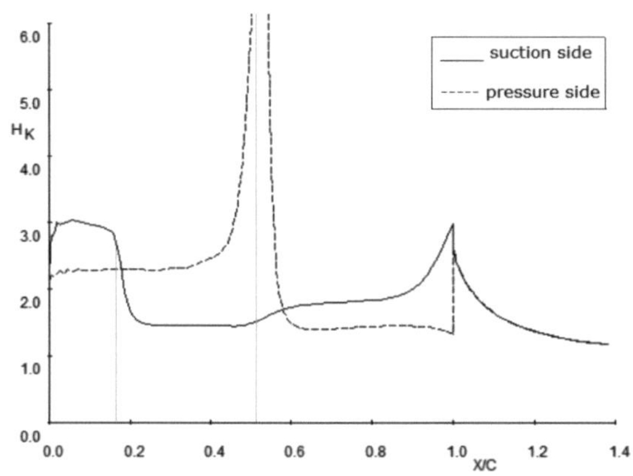

Figure 3.
Shape factor of boundary layer of airfoil S809 at AOA = 7°.

when the reference airfoil shows drastically decreased H values at $x/c \approx 0.2$. The drag from turbulent boundary layer is expected more at the reference airfoil in the suction side.

Moreover, the trailing edge of S809 suction side also has separation with high H value, while S809gx has the smaller H value at the trailing edge. The pressure side transition is thought to be more violent at the S809 by the extremely different H values of transition point of pressure side of each airfoil, see **Figures 2** and **3**, In detail, it can be seen from **Figures 2** and **3**, that the shape factor of the airfoil drops significantly at x/c≈3, while it is much earlier for the reference airfoil at x/c≈0.18. This indicates delayed transition location, which confirms the previous discussion. Interestingly, the level of H for both airfoils within the laminar region is comparable at around H ≈ 3.0, showing that the suction peak and its corresponding pressure gradient are comparable. A huge difference is observed on the pressure side when the transition occurs. Two main aspects can be seen: (1) transition occurs also at the delayed position at x/c≈0.65 compared to the reference airfoil at x/c≈0.5 and (2) the level of H at the location of transition is much smaller. The latter effect indicates that the laminar separation bubble is suppressed effectively for the optimized airfoil. This also indicates a reduced pressure gradient effect.

To check the validity of the optimized airfoil in soiled condition, boundary layer transition is forced to be 0.05 on the suction side and 0.1 on the pressure side, based on the roughness sensitivity experiment in [41], simulated in Rfoil for its 3D consideration [27].

Although GR values of airfoils are similar in different flow regimes in **Table 3**, the optimized one shows to have larger laminar boundary layer region over different AOA ranges in the forced transition situation, see **Figure 4**.

	S809gx	S809	S809gx	S809
AOA (°)	7	7	21.5	21.5
GR	61	61	7.53	6.67
Xtr (suction side)	0.05	0.05	0.0135	0.0025
Xtr (pressure side)	0.1	0.1	0.1	0.1

Table 3.
GR and Xtr values of airfoils with forced transition (Rfoil).

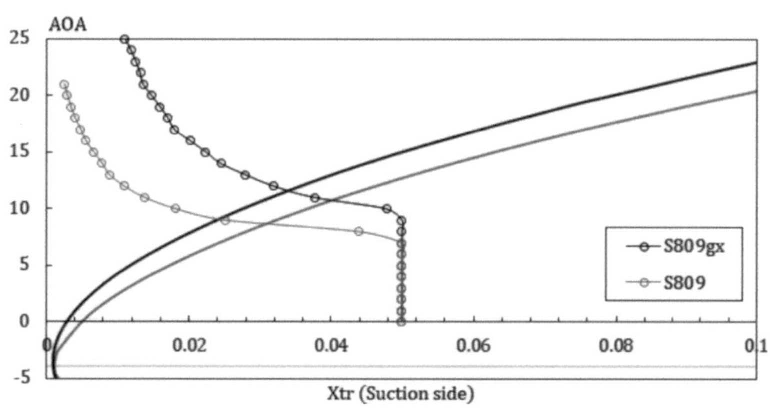

Figure 4.
Transition point (Xtr) over angle of attack.

Figure 2 shows the leading edge of airfoil as background of the graph. The airfoil S809gx has the larger laminar boundary layer region over all angle of attack values. It indicates that the optimized airfoil is shaped to have larger laminar boundary layer region even after transition is forced to be earlier than the normal state. Based on the assumption that the soiled condition triggers earlier boundary layer transition that occurs earlier than clean air condition [41], the optimized S809gx airfoil can be also useful under real air contamination situation [39], which will be discussed further in Section 5.

The wind turbine blade design with the optimized airfoil and the reference one is compared in **Figures 5** and **6**. The blade is designed with the same twist angle and chord length distribution based on Ref. [42], and the only difference is the airfoil type.

The blades designed with each airfoil are visualized in **Figures 5** and **6**. The airfoil distribution along the radial position with chord length distribution is based on the NREL Phase VI design guidelines [42], see **Tables 4** and **5**.

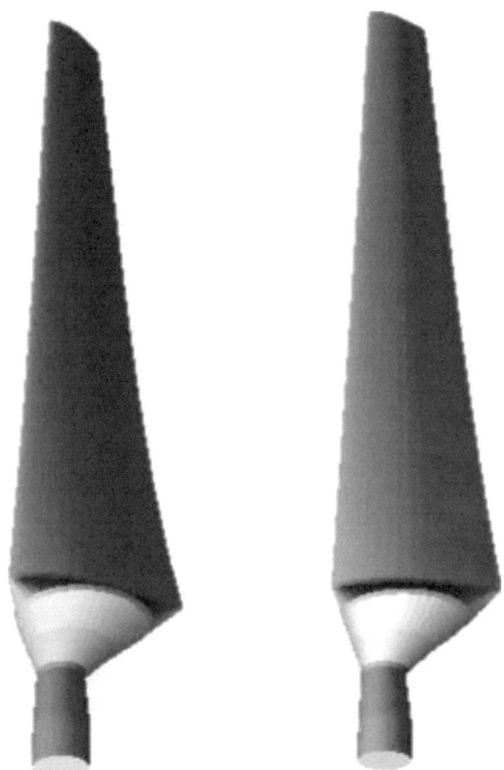

Figure 5.
Blade designed with the airfoil S809gx (left) and the airfoil S809 (right).

Figure 6.
Top view of the blade with S809gx (left) and S809 (right).

Radial position (m)	Chord length (m)	Twist (°)	Airfoil
0.508	0.218	0	Circular
0.66	0.218	0	Circular
0.883	0.183	0	Circular
1.008	0.349	0	Circular
1.067	0.441	0	Circular
1.133	0.544	0	Circular
1.257	0.737	20.04	S809
1.343	0.728	18.07	S809
1.51	0.711	14.29	S809
1.648	0.697	11.91	S809
1.952	0.666	7.98	S809
2.257	0.636	5.31	S809
2.343	0.627	4.71	S809
2.562	0.605	3.42	S809
2.867	0.574	2.08	S809
3.172	0.543	1.15	S809
3.185	0.542	1.115	S809
3.476	0.512	0.494	S809
3.781	0.482	−0.015	S809
4.023	0.457	−0.381	S809
4.086	0.451	−0.475	S809
4.391	0.42	−0.92	S809
4.696	0.389	−1.352	S809
4.78	0.381	−1.469	S809
5	0.358	−1.775	S809

Table 4.
Blade property of NREL Phase VI.

Radial position (m)	Chord length (m)	Twist (°)	Airfoil name
0.508	0.218	−3.00	Circular
0.660	0.218	−3.00	Circular
0.883	0.183	−3.00	Circular
1.008	0.349	−3.00	Circular
1.067	0.441	−3.00	Circular
1.133	0.544	−3.00	Circular
1.257	0.737	17.04	s809gx
1.343	0.728	15.07	s809gx
1.510	0.711	11.29	s809gx
1.648	0.697	8.91	s809gx
1.952	0.666	4.98	s809gx
2.257	0.636	2.31	s809gx

Radial position (m)	Chord length (m)	Twist (°)	Airfoil name
2.343	0.627	1.71	s809gx
2.562	0.605	0.42	s809gx
2.867	0.574	−0.92	s809gx
3.172	0.543	−1.85	s809gx
3.185	0.542	−1.89	s809gx
3.476	0.512	−2.51	s809gx
3.781	0.482	−3.02	s809gx
4.023	0.457	−3.38	s809gx
4.086	0.451	−3.475	s809gx
4.391	0.420	−3.92	s809gx
4.696	0.389	−4.35	s809gx
4.780	0.381	−4.47	s809gx
5.000	0.358	−4.78	s809gx

Table 5.
Airfoil S809gx distribution along the blade.

3. Aerodynamic parameters

Regarding the aerodynamic parameters like Cl, Cd, and GR, two airfoils show similar distribution over the angles of attack. However, the optimized airfoil indicates slightly increased Cl and decreased Cd. Those small advantageous differences are summed up to show increased GR.

The Cl values are corrected with the 3D correction law, mainly considering twist angle and chord per radius ratio of the blade in turbine unit. The corrected value conveys the effect of the Coriolis, centrifugal force, delay of separation, and so on in rotational augmentation [32, 33]. As the rotational effect is significant in lift force, the correction law is only applied in Cl, not Cd. The reference experiments are found in the works of [45, 46]. The calculations are done in $Re=10^6$.

The optimized airfoil Cl values show slightly advantageous over stall angle of attack region compared to the reference one, see **Figure 7**. The drag coefficient is also smaller than the reference, as shown in **Figure 8**. Although the airfoil was designed to have better GR value by 2D calculation at the target of angle of attack of 7°,

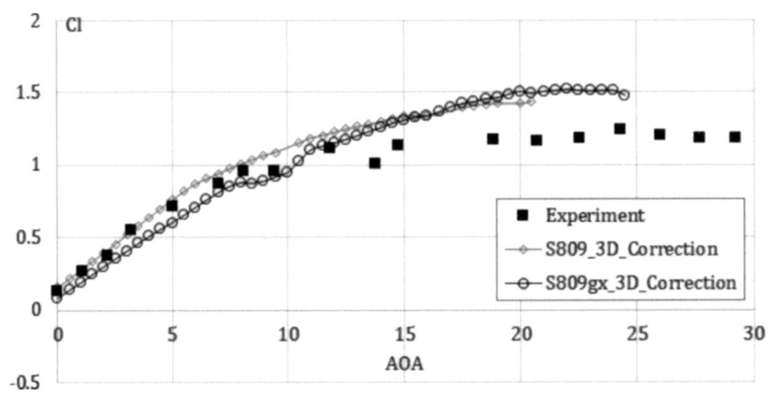

Figure 7.
Cl distribution over angle of attack (°).

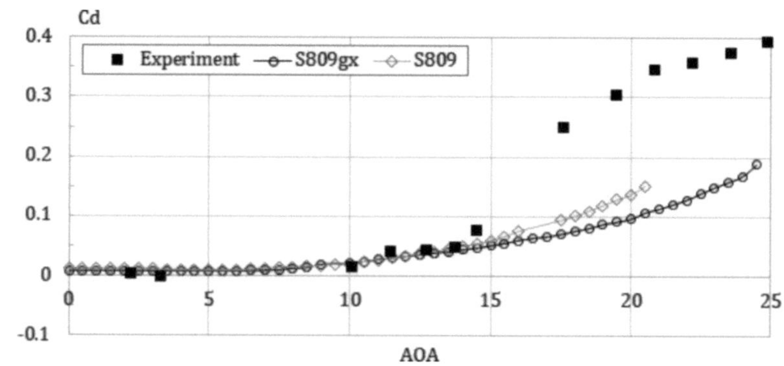

Figure 8.
Cd distribution over angle of attack (°).

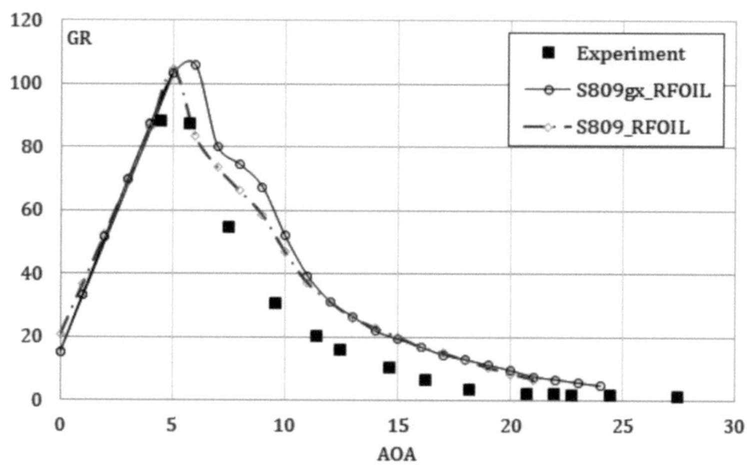

Figure 9.
GR vs. AOA (°).

the 3D corrected value also shows the advantage of airfoil S809gx over the different angle of attack ranges.

Cl values corrected by 3D correction law of Hansen [31] in **Figure 7** are also supported by the GR calculation of software Rfoil, which improves the 2D prediction with the treated laminar and turbulent boundary layer closure problems [47, 48] in **Figure 9**. Based on the Rfoil validity [27], increased GR values of the airfoil S809gx especially at fully attached angle of attack range (5–13°) show positive implication for improvement of the following power production in the wind turbine unit.

4. Thrust and power curves

To run the code simulation for reference turbine and turbine with optimized airfoil, operation properties are set as **Table 6**. A fixed pitch value of 3° (equal to turbine 1) is controlled in blade distribution property in **Table 5** for turbine 2. Blade 1 is designed with the reference airfoil S809 [42], and blade 2 is designed with the optimized airfoil S809gx.

Although the advantage of the optimized airfoil in Cl, Cd, and GR values seems to be negligible in **Figures 7–9**, power production curve shows how blade lift efficiency is improved by those airfoil construction, as shown in **Figure 10**. The thrust

	Turbine 1	Turbine 2
Power regulation	Stall	Stall
Transmission	Single	Single
V cut in (m/s)	6.00	6.00
V cut out (m/s)	25.00	25.00
Rotational speed (rpm)	71.63	71.63
Outer radius (mm)	5000.00	5000.00
Fixed pitch (°)	3.00	0.00
Variable losses	0.22	0.22
Blade type	Blade 1	Blade 2

Table 6.
Properties for turbines [42].

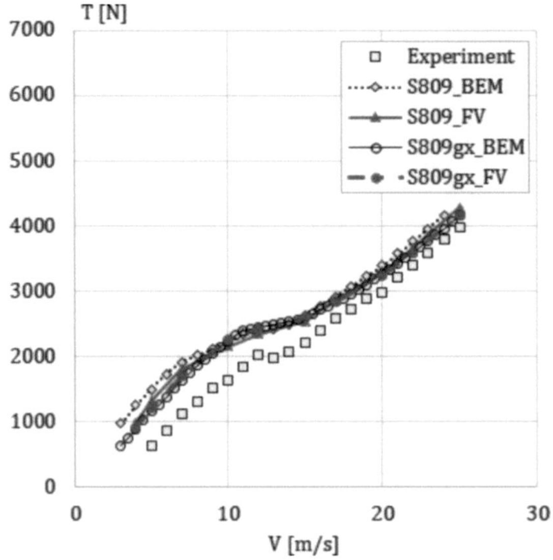

Figure 10.
Power-velocity curve.

forces are calculated to show the values that are similar at blades with both airfoils, see **Figure 11**. The similarity in value distribution of thrust implies the power increment that is caused by increased lift velocity from the blade designed with the optimized airfoil.

As the Power Production is calculated by all lifting efficiencies of each blade section [48], combined advantage of each section of blade airfoil produces largely increased power production simulation, especially in the inflow velocity range of 7–25 m/s. Considering the discrepancies between different tools of calculation, the optimized airfoil turbine produced 150% larger power than the reference one. As the optimized airfoil turbine power prediction is based on the simulation of reference one, validated with its experimental data, the discrepancy between experimental data and calculation in high velocity (15–25 m/s) should be considered more.

The thrust curves show similar value pattern with B-Go code, except the fact that thrust force is estimated to be higher at the velocity of flow stall regime, where the prediction can be misled in BEM and FV codes [38]. The B-Go code thrust calculation also supports the increment in blade velocity with the optimized airfoil as the thrust is not increased drastically in the turbine blade designed with the S809gx, see **Figure 12**.

The power value validation in high velocity, which would cause stall delay in blade [33], is tried with the state-of-the-art-code B-Go. The B-Go has been coded with the off design flow region where separation frequently occurs to make BEM code to be

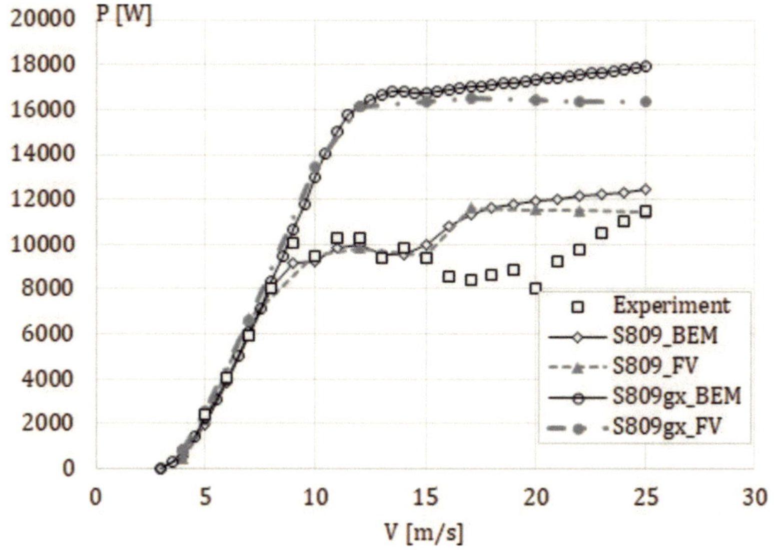

Figure 11.
Thrust-velocity curve.

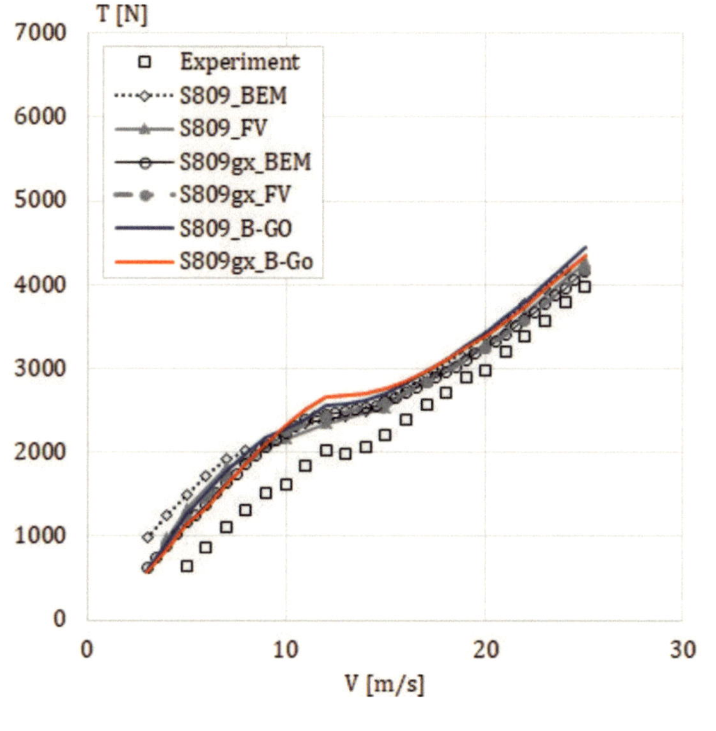

Figure 12.
Thrust-velocity curve with B-GO code.

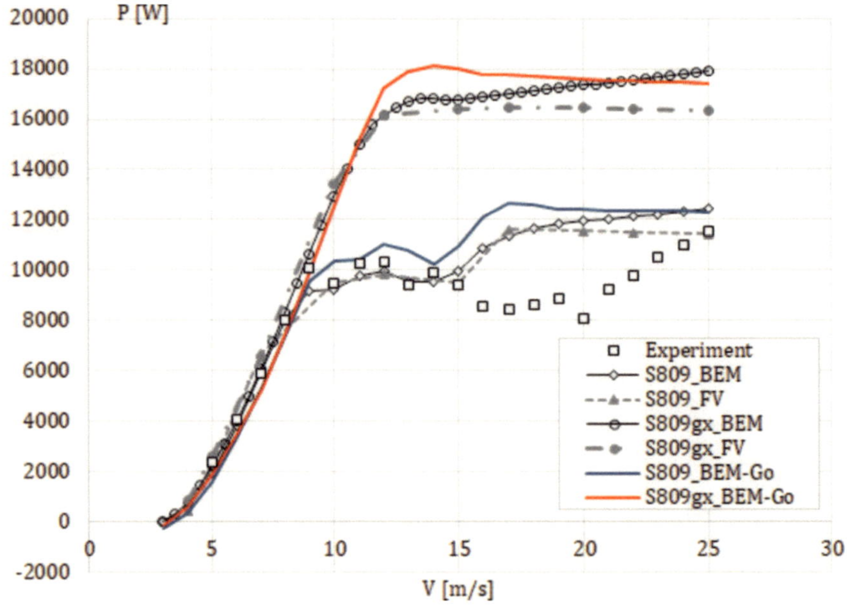

Figure 13.
Power-velocity curve with B-GO code.

challenged in prediction [38], the power values at high velocity are predicted in spite of the discrepancy with experimental data. Although BEM and FV calculations show the value gap in the velocity of 15–25 m/s region, the B-GO codes show the similar pattern with the other codes. The increasing power values in stall region by BEM prediction show the weakness of BEM at the stall region [37, 38]. It also implies the possibility, which experimental data might have had the error in the stall region, as the experimental set up also has their limit in stall region measurements. The optimized airfoil turbine shows c.a. 150% higher power production in stall region, see **Figure 13**.

5. Performance of the optimized airfoil under soiled condition

As it is shown in Section 2, the optimized airfoil shape shows larger laminar boundary layer regimes even under a forced transition situation, which imitates the soiled condition, as shown in **Figure 4**. Not only smaller drag within the boundary layer but also the lift coefficient at the forced transition is benefitted because of the enlarged laminar boundary region at optimized airfoil. As it is shown in **Figure 14**, the Cl value difference of optimized airfoil at forced transition and normal transition situation is negligible. The influence of difference of Cl on GR, calculated by Rfoil, is also demonstrated in **Figure 15**. The optimized airfoil is less sensitive to changes in inflow conditions. This is not only caused by its 2D characteristics but also being supported by the 3D rotational effects, which delays flow separation and reduces the turbulent boundary layer drag [32, 33]. The optimized airfoil can be tolerable in efficiency decrement in soiled condition or other causes of earlier transition occurrences.

6. Conclusions

Stochastic optimization, GA, has been applied to optimize airfoil shape toward larger GR and advantageous boundary layer transition in HAWT. The resulted

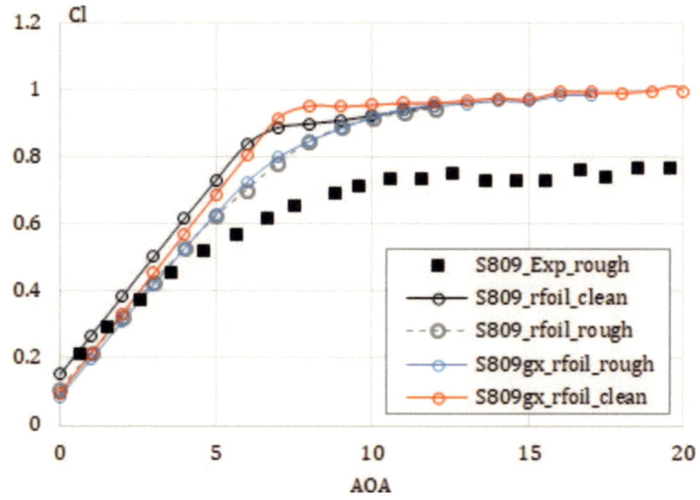

Figure 14.
Cl distribution at soil condition with experiment [49].

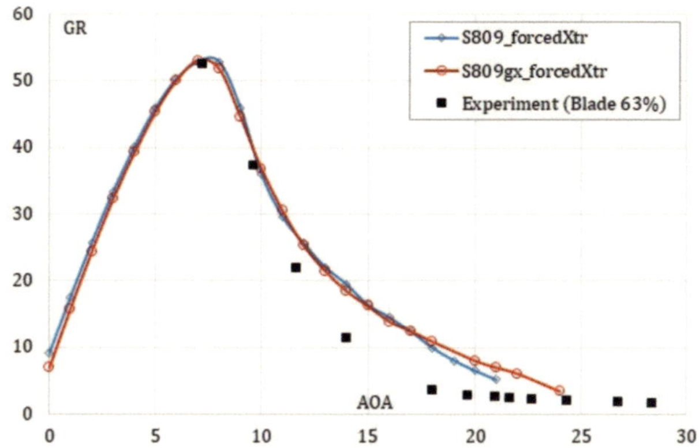

Figure 15.
GR comparison of airfoil in forced transition with experiment [42].

airfoil shows a 121% higher GR, c.a. 120–170% larger laminar boundary layer on the airfoil surface in targeted AOA (°). The Cl, Cd, and GR values of two airfoils seem to be slight in the results; however, the power production predicted by different codes shows the combined effect of optimized airfoil rotor sections that lead to 150% higher power production. The thrust curves show a similar distribution pattern, indicating the velocity of the blade designed with optimized airfoil influences the power improvement, not the thrust force. The corrected BEM code with 3D rotational augmentation and B-Go codes for the stall region are used for compensating the prediction weakness of BEM in flow separation. The airfoil validity in soil condition is simulated with a forced transition, which shows a negligible lifting coefficient decrement in the optimized airfoil. The laminar boundary layer is still broader at optimized airfoil in forced transition, which indicates that the optimized airfoil shape is useful for realistic airflow with dirt and 3D rotation.

Acknowledgements

Busan Brain 21 project of BMC is appreciated for its funding in this research.

Author details

Youjin Kim[1*], Galih Bangga[2] and Antonio Delgado[1]

1 Institute of Fluid Mechanics, FAU Erlangen-Nürnberg, Erlangen, Germany

2 Institute of Aerodynamics and Gas Dynamics, University of Stuttgart, Stuttgart, Germany

*Address all correspondence to: youjin.kim@fau.de

References

[1] Jebaraj S, Iniyan S. A review of energy models. Renewable and Sustainable Energy Reviews. 2006;**10**(4):281-311. DOI: 10.1016/j.rser.2004.09.004

[2] Baños R, Manzano-Agugliaro F, Montoya F, Gil C, Alcayde A, Gómez J. Optimization methods applied to renewable and sustainable energy. Renewable and Sustainable Energy Reviews. 2011;**15**(4):1753-1766

[3] Lei M, Shiyan L, Chuanwen J, Hongling L, Yan Z. A review on the forecasting of wind speed and generated power. Renewable and Sustainable Energy Reviews. 2009;**13**(4):915-920

[4] Miller A, Chang B, Issa R, Chen G. Review of computer-aided numerical simulation in wind energy. Renewable and Sustainable Energy Reviews. 2013;**25**:122-134

[5] Serrano González J, Burgos Payán M, Santos JMR, González-Longatt F. A review and recent developments in the optimal wind-turbine micro-siting problem. Renewable and Sustainable Energy Reviews. 2014;**30**:133-144

[6] Fuglsang P, Aagaard Madsen H. Optimization of stall regulated rotors. In: Proceedings of 1995 American Society of Mechanical Engineers (ASME) Energy Sources Technology Conference and Exhibition. Houston, TX, United States; 1995

[7] Fuglsang P, Aagaard Madsen H. A Design Study of a 1 MW Stall Regulated Rotor. Roskilde, Denmark: Risø National Laboratory; 1995

[8] Fuglsang P, Bak C, Schepers J, Bulder B, Cockerill T, Claiden P, et al. Site specific design optimization of wind turbines of 1.5-2.0MW wind turbines. Wind Energy. 2002;**5**(4):261-279

[9] Bak C. Aerodynamic design of wind turbine rotors. In: Brønsted P, Nijssen R, editors. Advances in Wind Turbine Blade Design and Materials. Sawston, Cambridge: Woodhead Publishimg Limited; 2013. ISBN 978-1-84569-580-4

[10] Ning A, Damiani R, Moriarty P. Objectives and constraints for wind turbine optimization. In: 31st ASME Wind Energy Symposium. 2013

[11] Fuglsang P, Madsen HA. Optimization method for wind turbine rotors. Journal of Wind Engineering and Industrial Aerodynamics. 1999;**80**:191-206

[12] Xudong W, Shen WZ, Zhu WJ, Sorensen JN, Jin C. Shape optimization of wind turbine blades. Wind Energy. 2009;**12**:781-803

[13] Maki K, Sbragio R, Vlahopoulos N. System design of a wind turbine using a multi-level optimization approach. Renewable Energy. 2012;**43**:101-110

[14] Benini E, Toffolo A. Optimal design of horizontal-axis wind turbines using blade-element theory and evolutionary computation. ASME Journal of Solar Energy Engineering. 2002;**124**:357-363

[15] Eke G, Onyewudiala J. Optimization of wind turbine blades using genetic algorithm. Global Journal of Researches in Engineering. 2010;**10**

[16] Morgan C, Garrad A. The design of optimum rotors for horizontal axis wind turbines. In: Proceedings of the 10th BWEA Wind Energy Conversion Conference. London, England; 1988

[17] Snel H. Review of aerodynamics for wind turbines. Wind Energy. 2003;**6**:203-211

[18] Li JY, Li R, Gao Y, Huang J. Aerodynamic optimization of wind

turbine airfoils using response surface techniques. Proceedings of the Institution of Mechanical Engineering. 2010;**224**:827-838

[19] Liao CC, Zhao XL, Xu JZ. Blade layers optimization of wind turbines using FAST and improved PSO algorithm. Renewable Energy. 2012;**42**:227-233

[20] Wang L, Wang TG, Luo Y. Improved non-dominated sorting genetic algorithm (NSGA)-II in multi-objective optimization studies of wind turbine blades. Applied Mathematics and Mechanics. 2011;**32**:739. DOI: 10.1007/s10483-011-1453-x

[21] Chehouri A, Younes R, Ilinca A, Perron J. Review of performance optimization techniques applied to wind turbines. Applied Energy. 2015;**142**:361-388

[22] Fogel DB. Introduction. In: Bäck T, Fogel DB, Michalewicz Z, editors. Handbook of Evolutionary Computation. San Francisco, California: Morgan Kaufmann Publisher; 1997. pp. 1-2. ISBN: 0750303921

[23] Windle BCA. Mendel and his theory of heredity. In: A Century of Scientific Thought and Other Essays. Burns & Oates; 1915

[24] King RC et al. A Dictionary of Genetics. 7th ed. Oxford, England: University Press; 2006. p. 129. ISBN: 978-0-19-530761-0

[25] Kim Y, Al-Abadi A, Delgado A. Airfoil Boundary Layer Optimization Toward Aerodynamic Efficiency of Wind Turbines: Chapter from the Book Flight Physics—Models, Techniques and Technologies

[26] Snel H, Houwink R, Bosschers J, Piers WJ, Bruining A. Sectional Prediction of 3-D Effects for Stalled Flows on Rotating Blades and Comparison with Measurements: Proceedings of the ECWEC '93 Conference. Travemunde, Germany. pp. 395-399

[27] van Rooij, RPJOM. Modification of the Boundary Layer Calculation in RFOIL for Improved Airfoil Stall Predictio: Report IW-96087R, Delft University of Technology, Delft, The Netherlands. Available from: http://www.windenergy.citg.tudelft.nl/

[28] Himmelskamp H. Profile Investigation on a Rotating Airscrew, Technical Report [dissertation], Gottingen, Germany; 1945

[29] Ronsten G. Static Pressure Measurements on a Rotating and a Non-Rotating 2.375 m Wind Turbine Blade—Comparison with 2-D Calculations, Proceedings of the EWEC '91 Conference. Amsterdam; 1991. pp. 214-220

[30] Bruining A, Van Bussel GJW, Corten GP, Timmer WA. Pressure Distribution from a Wind Turbine Blade; Field Measurements Compared to 2-Dimensional Wind Tunnel Data, Technical Report IW-93065R. Delft University of Technology, Institute of Wind Energy; 1993

[31] Hansen MOL, Soerensen JN, Michelsen JA, Soerensen NN. A Global Navier-Stokes Rotor Prediction Model, 35th AIAA Aerospace Sciences Meeting & Exhibit. Reno: AIAA paper; 1997. pp. 97-0970

[32] Snel H, Houwink R, Piers WJ. Sectional Prediction of 3-D Effects for Separated Flow on Rotating Blades. Avignon, France: Eighteenth European Rotorcraft Forum; 1992

[33] Snel H, Houwink R, Bosschers J, Piers WJ, Bruining A. Sectional Prediction of 3-D Effects for Stalled Flows on Rotating Blades and Comparison with Measurements,

Proceedings of the ECWEC '93 Conference. Travemunde, Germany; 1993. pp. 395-399

[34] Soerensen DN, Soerensen JN. Quasi 3-dimensional model for rotating airfoil. In: 13th AIAA Applied Aerod Conference. San Diego, CA: AIAA; 1995. pp. 95-1916

[35] Shen WZ, Soerensen JN. Quasi-3D Navier-stokes model for rotating airfoil. Journal of Computational Physics. 1999;**150**:518-548

[36] Ingram G. Wind Turbine Blade Analysis Using the Blade Element Momentum Method Version 1.0. Durham, UK: School of Engineering, Durham University; 2005

[37] van Garrel A. Development of a wind turbine aerodynamics simulation module, ECN-C-03-079; 2003

[38] Bangga G. Comparison of blade element method and CFD simulations of a 10MW wind turbine. Fluids. 2018;**3**(4):73. DOI: 10.3390/fluids3040073

[39] van Rooij RPJOM, Timmer WA. Roughness Sensitivity Consideration for Thick Rotor Blade Airfoils, ASME 2003. Vol. 1252003

[40] Corten GP, Veldkamp HF. Insects cause double stall. In: European Wind Energy Conference. Copenhagen, Denmark; 2001. pp. 470-474

[41] Fuglsang P, Antoniou I, Dahl KS, Aagaard Madsen H. Wind Tunnel Tests of the FFA-W3-241, FFA-W3-301 and NACA 63-430 Airfoils. Denmark: Forskningscenter Risoe. Risoe-R, No. 1041(EN); 1998

[42] Hand MM, Simms DA, Fingersh LJ, Jager DW, Cotrell JR, Schreck S, Larwood SM. Unsteady Aerodynamics Experiment Phase VI: Wind Tunnel Test Configurations and Available Data Campaigns. National Renewable Energy Laboratory. NREL/TP-500-29955; 2001

[43] Giguère P, Selig MS. Design of a Tapered and Twisted Blade for the NREL Combined Experiment Rotor, Department of Aeronautical and Astronautical Engineering University of Illinois at Urbana-Champaign Urbana, Illinois, NREL. NREL/SR-500-26173; 1999

[44] Schlichting H. Boundary-Layer Theory. 9th ed. NewYork, USA: Springer; 1979

[45] Breton S-P, Coton FN, Moe G. A study on rotational effects and different stall delay models using a prescribed wake vortex scheme and NREL phase VI experiment data. Wind Energy. 2008; **11, 5**:459-482

[46] Lindenburg C. Investigation into Rotor Blade Aerodynamics Analysis of the stationary measurements on the UAE phase-VI rotor in the NASA-Ames wind tunnel. ECN-C--03-025; 2003

[47] Kevorkian J, Cole JD. Multiple Scale and Singular Perturbation Methods. Applied Mathematical Sciences. Vol. 114. New York: Springer; 1996

[48] Manwell JF, McGowan JG. Chapter 3. In: Wind Energy Explained. 2nd Eidition ed. WILEY; 2009

[49] Musial WD, Butterfield CP, Jenks MD. A Comparison of Two- and Three- Dimensional S809 Airfoil Properties for Rough and Smooth HAWT Rotor Operation, 9th ASME Wind Energy Symposium New Orleans, Louisiana, Solar Energy Research Institute (SERI), SERI/TP-257-3603 UC Category: 261 DE89009512

Chapter 4

Structural Optimization of Wind Turbine Blades for Improved Dynamic Performance

Gerges Edwar Mehanny Beshay and Karam Yousef Maalawi

Abstract

The design of the main structure of a wind turbine blade is optimized aiming at the improvement of the overall dynamic performance. Three optimization strategies are developed and tested. The first fundamental one is based on minimizing the total structural mass of the blade spar under frequency and strength constraints. The second and third strategies are concerned with the reduction of the overall vibration level by either minimizing a frequency-placement index or maximizing the natural frequencies and placing them at their target values to avoid large amplitudes and resonance occurrence. Design variables include cross-sectional dimensions and material properties along the spanwise direction of the blade spar. The optimization problem is formulated as a nonlinear constrained problem solved by sequential quadratic programming (SQP) technique. Two specific layup configurations, namely, circumferentially asymmetric stiffness (***CAS***) and circumferentially uniform stiffness (***CUS***), are analyzed. Exact analytical methods are applied to calculate the natural modes of vibration of a composite, thin-walled, tapered blade spar. The influence of coupling on the vibration modes is identified, and the functional behavior of the frequencies with the lamination parameters is thoroughly investigated and discussed. Finite element modeling using NX Nastran solver is performed in order to validate the analytical results. As a case study, optimized blade spar designs of a 750-kW horizontal axis wind turbine are given. The attained solutions show that the approach used in this study enhances the dynamic characteristics of the optimized spar structures as compared with a known baseline design of the wind turbine blade.

Keywords: wind turbine blades, structural optimization, natural frequencies, advanced composites, sequential quadratic programming, finite element method

1. Introduction

Among all renewable energies of different styles, wind energy is the most popularized and potentially applicable type of green energy. Because larger wind turbines have more power capture and economic advantages, the typical size of utility-scale wind turbines, as shown in **Figure 1**, has grown dramatically over the last three decades [1, 2]. Such large flexible configuration, operating in uncertain environments, gives rise to significant vibration problems and assesses the

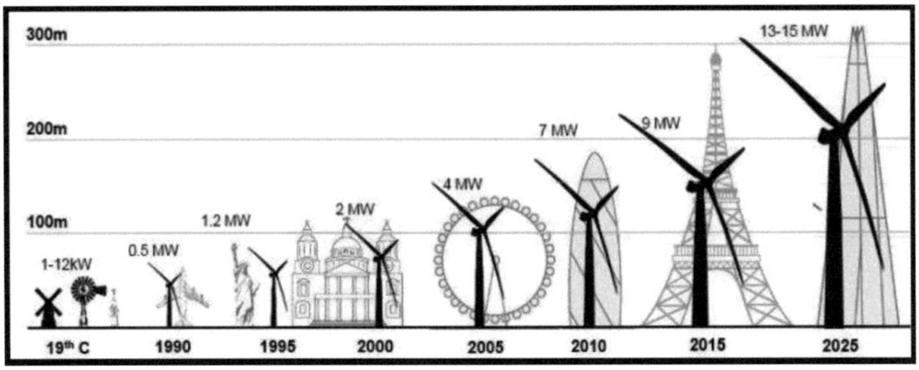

Figure 1.
Size and power increase of commercial wind turbines.

importance of analyzing structural dynamics in the design of successful wind turbine systems. The main supporting structures of the rotating blades are usually fabricated from thin-walled composite beams with a variety of cross-sectional types. These configurations are used extensively in the design of many aerodynamic structures because of their light weight-to-stiffness ratio and long fatigue life. To design these components, the dynamic characteristics, especially near-resonant conditions, need to be well examined to assure a safe operation.

The objective of this investigation is to optimize the structural dynamics of a thin-walled composite blade through the minimization of structural mass or reduction of the overall vibration level. The latter can be attained directly by maximizing the natural frequencies of the main blade structure under strength and mass constraints. In general conditions, however, we need a material that is as light as possible for a specified stiffness in order to satisfy the design criteria and to minimize the weight-induced fatigue loads.

The main advantages of fiber composite materials [3] are their high strength and stiffness combined with low density, their superior fatigue properties due to the prevention of crack propagation, and their ability to tailor the layup for optimum strength and stiffness. However, sharp transitions between component materials may cause stress and strain discontinuities that facilitate failure [4]. A solution that can be promising to enhance dynamics and aeroelastic stability of composite blades is the use of functionally graded materials (FGMs), in which the mechanical and physical properties vary spatially within the structure. The concept of functionally graded materials was originated in Japan in 1984 during a space project, in the form of proposed thermal barrier material capable of withstanding high-temperature gradients [5]. *FGMs* may also be developed using fiber-reinforced layers with a changing volume fraction of fibers, rather than constant, producing grading of the material with favorable properties [6].

Considering, next, optimization of wind turbine blades, Maalawi and Negm [7] presented an optimization model for the design of a typical blade structure of horizontal axis wind turbines. The main blade spar was represented by thin-walled tubular beam composed of uniform segments with the design variables chosen to be the cross-sectional area, radius of gyration, and length of each segment. The optimal design is pursued with respect to maximum frequency design criterion subject to mass and aeroelastic constraints. The optimization problem was solved by multidimensional search techniques, where the aeroelastic stability boundaries and steady-state response were calculated using Floquet's transition matrix theory. Another work by Maalawi [8] developed an optimization model for placing the frequencies of a wind turbine tower/nacelle/rotor structure in free yawing motion. The mathematical formulation considered a single pole tower configuration having thin-walled circular cross section with constant taper along the tower height. The nacelle/rotor combination was modeled as a rigid mass elastically supported at the top of the tower by the torsional spring of the yawing mechanism. The resulting governing differential equation of motion was solved analytically by transforming it into a standard form of Bessel's equation, which leads to the necessary exact solutions for the frequencies and mode shapes. Useful design charts were developed for placing the frequencies at their needed target values with no penalty of increasing the total structural weight of the system.

In the context of using the concept of material grading, Librescu and Maalawi [9] formulated an analytical approach for obtaining the optimal design of a class of solid nonuniform composite wings with improved aeroelastic stability. The objective function was measured by maximizing the wing divergence speed while maintaining the total structural mass at a value equal to that of a known baseline design. Exact solutions were obtained for different categories of unidirectionally reinforced composite wing structures: namely, the linear volume fraction (*L-VF*), the parabolic volume fraction (*PR-VF*), and the piecewise volume fraction (*PW-VF*) wing models. Results revealed that in general, the torsional stability of the wing can be substantially improved by using nonuniform, functionally graded composites instead of the traditional ones having uniform volume fractions of the constituent materials.

The optimization of large wind turbines was considered by Kun-Nan Chen and Pin-Yung Chen [10], who applied a two-step procedure for finding the optimum design of composite blades. The first step concerned the optimal aerodynamic shape of the blade as described by the chord and twist angle distributions in the spanwise direction. The second step yielded the optimal material distribution. A 3-MW wind turbine with blades having cross sections of NREL S818, S825, and S826 airfoil types is demonstrated as a case study. A parameterized finite element model of the aerodynamically optimized blade was created using the ANSYS software. The optimization results showed that the initial blade model is an infeasible design due to a high level of the maximum stress, exceeding the upper limit of the stress constraint, but eventually the process converges to a feasible solution with the expense of increased total mass of the blade. Another work by Maalawi and Badr [11] considered the excessive wind turbine blade vibrations induced by continuous pitching, which is necessary to limit the power output and protect the generator from damages in severe wind conditions. They utilized analytical Bessel's functions of the first kind which yield to the exact solutions of the resulting governing differential equation. The associated optimization problem was formulated by considering two forms of the objective function. The first one was represented by a direct maximization of the fundamental frequency, while the second one considered minimization of the square of the difference between the fundamental frequency and its target or desired value. In both strategies, an equality constraint is imposed on the total structural mass in order not to violate other economic and performance requirements. Design variables encompass the blade tapering ratio, chord, and shear wall thickness distributions. Danny Sale et al. [12] developed a numerical methodology for the structural analysis and optimization of composite blades for wind and hydrokinetic turbines. They derived a structural mechanic model which is based upon a combination of classical lamination theory with a Euler-Bernoulli and shear flow theory applied to composite beams. The development of this simplified structural model was motivated by the need for an accurate and computationally efficient method that is suitable for parametric design and optimization studies of composite blades. An important characteristic of this structural model is its ability to handle complex geometric shapes and isotropic or anisotropic composite layups. For a specified design load, the objective of the structural optimization was to minimize the blade's mass while satisfying constraints on maximum allowable stress, blade tip deflection, buckling, and placement of blade natural frequencies. Adam Chehouri et al. [13] presented an improved version of the preliminary optimization tool called CoBlade, which offers designers and engineers an accelerated design phase by providing the capabilities to rapidly evaluate alternative composite layups and study their effects on static failure and fatigue of wind turbine blades. In this study, the optimization formulations included nonlinear failure constraints, and a comparison between three formulations was made to show the importance of choosing the blade mass as the main objective function and the inclusion of failure constraints in the wind turbine blade design.

A recent work by Maalawi [14] introduced a mathematical model for optimizing dynamic performance of thin-walled functionally graded box beams with closed cross sections. The objective function was measured by maximizing the natural frequencies and places them at their target values to avoid the occurrence of large amplitudes of vibration. Variables include fiber volume fraction, fiber orientation angle, and ply thickness distributions. Various power law expressions describing the distribution of the fiber volume fraction have been implemented, where the power exponent was taken as a main optimization variable. The mass of the structure is kept at a value equal to that of a known reference beam. Side constraints were also imposed on the design variables in order to avoid having unacceptable optimal solutions. A case study, including optimization of a cantilevered, a single-cell spar beam made of carbon/epoxy composite was considered. Conspicuous design charts were developed, showing the optimum design trends for the mathematical models implemented in the study. It was shown that the developed mathematical models are adequately satisfying the required global optimization of typical composite, functionally graded, thin-walled beam structures.

This chapter focuses on the optimization of the main structure of a wind turbine blade by either minimizing structural mass under frequency and strength constraints or maximizing the natural frequencies under mass, strength, and side constraints. This model is applied to tapered, anisotropic spar beam with thin-walled closed cross section made of laminated fibrous composites with variable thickness and stiffness. The study is focused on the spar structure that represents the main load-carrying component of the wind turbine blade. Material grading concept is utilized by changing the fiber content throughout the blade structure. Design variables include the volume fraction distribution of the constituent materials of construction and geometric and cross-sectional parameters of the blade spar. The blade is assumed to have a large span-to-chord ratio, which enabled us to model the main spar as an equivalent straight beam, positioned along the elastic axis. Structural analyses are performed using simplified mathematical expressions by implementing the conventional beam and classical lamination theories. The governing differential equations of motion are derived and solved by the transfer-matrix method for the coupled extensional-torsional and flexural-torsional modes of vibration. A case study is given considering thin-walled blade spar of a 750-KW horizontal axis wind turbine. Numerical results are presented and discussed showing the success of the developed mathematical model in producing efficient blade designs with improved dynamic performance. Finally, the relevant concluding remarks and recommendations for future studies are given and discussed.

2. Governing differential equations of motion

Among the dynamic characteristics of the blade main structures, determination of the natural frequencies and the associated mode shapes is of fundamental importance. An analytical model for the free vibration of anisotropic thin-walled beams with closed cross sections was developed by Armanios and Badir [15] using a variational asymptotic approach and Hamilton's principles. This model was applied to arbitrary closed cross sections made of laminated fibrous composites with variable thickness and stiffness. The analysis was applied to two kinds of laminated composites: circumferentially uniform stiffness (CUS) and circumferentially asymmetric stiffness (CAS). The model was also implemented in Refs. [16, 17] to investigate the influence of coupling on the free vibration of thin-walled composite beams. Shadmehri et al. [18] studied the static and dynamic characteristics of composite thin-walled beams that are constructed from a single-cell cross section. The structural model considered incorporated a number of nonclassical effects, such as material anisotropy, transverse shear, warping inhibition, nonuniform torsion, and rotary inertia. The governing equations were derived using extended Hamilton's principle and solved using extended Galerkin's method. Phuong and Lee [19] presented a flexural-torsional analysis of thin-walled composite box beams. A general analytical model applicable to thin-walled composite box beams subjected to vertical and torsional loads was developed. Analytical solutions for the free vibration analysis of tapered thin-walled laminated composite beams with closed cross sections are given in Ref. [20]. The exact values of frequencies were obtained by means of power series schemes. A parametric analysis was performed for different taper ratios, stacking sequences, and materials.

Considering functionally graded constructions, Kargarnovin and Hashemi [21] investigated the free vibration of a fiber composite cylinder, in which the volume fraction of fibers varies longitudinally, using a semi-analytical method. The distribution of volume fraction of fiber in base matrix was based on power law model. Another study by Liu and Shu [22] developed an analytical solution to study the free vibration of exponential functionally graded beams with a single delamination. They showed that the natural frequencies increase as Young's modulus ratio of the constituent materials becomes bigger.

Figure 2 shows the structural model of the blade spar, which is represented by a thin-walled cantilever beam which consists of $\boldsymbol{Ns}$ uniform segments. Each segment has different dimensions and material properties that satisfy the geometrical tapering and material grading distribution. Any segment $\boldsymbol{k}$ with length $\boldsymbol{L_k}$ has a rectangular cross section with dimensions, width $\boldsymbol{b_k}$, depth $\boldsymbol{a_k}$, and wall thickness $\boldsymbol{H_k}$. Each segment is a uniform laminated fibrous composite beam which consists of $\boldsymbol{N_r}$ layers, each of which has thickness $\boldsymbol{h_j}$, fiber volume fraction $\boldsymbol{V_{fj}}$, and fiber orientation angle $\boldsymbol{\theta_j}(j = 1,2, \dots, Nr)$.

The constitutive relationships in terms of stress resultants and kinematic variables are [15, 16]:

$$\begin{bmatrix} T \\ M_x \\ M_y \\ M_z \end{bmatrix} = \begin{bmatrix} C_{11} & C_{12} & C_{13} & C_{14} \\ C_{12} & C_{22} & C_{23} & C_{24} \\ C_{13} & C_{23} & C_{33} & C_{34} \\ C_{14} & C_{24} & C_{34} & C_{44} \end{bmatrix} \begin{bmatrix} U_1' \\ \varphi' \\ U_3'' \\ U_2'' \end{bmatrix} \tag{1}$$

where T is the tensile force, M_x is the torsional moment, and M_y and M_z are the bending moments about the y and z axes, respectively. C_{mn} are called the beam cross-sectional stiffness coefficients, and U_1, U_2, and U_3 are the average cross-sectional displacements along x, y, and z coordinates, respectively, and $\varphi(x)$ is the elastic twist about the x axis. The prime denotes differentiation with respect to x. Applying Hamilton's principle, the equations of undamped free vibration are [15]:

$$\begin{aligned}
&C_{11}U_1'' + C_{12}\varphi'' + C_{13}U_3''' + C_{14}U_2''' - m\acute{U}_1 = 0 \\
&C_{12}U_1'' + C_{22}\varphi'' + C_{23}U_3''' + C_{24}U_2''' - I\acute{\varphi} - S_z\acute{U}_3 + S_y\acute{U}_2 = 0 \\
&C_{13}U_1''' + C_{23}\varphi''' + C_{33}U_3'''' + C_{34}U_2'''' + S_z\acute{\varphi} + m\acute{U}_3 = 0 \\
&C_{14}U_1''' + C_{24}\varphi''' + C_{34}U_3'''' + C_{44}U_2'''' - S_y\acute{\varphi} + m\acute{U}_2 = 0
\end{aligned} \tag{2}$$

where m, I, and S_z and S_y are the mass, polar, and first moments of inertia per unit length of the beam, respectively. The dot superscript denotes differentiation with respect to time.

A closed-form solution for the most general case of the equations of motion (Eq. 2) is not available. Two particular cases of fiber layup are considered in which some of the stiffness coefficients vanish. The first case is called circumferentially uniform stiffness (CUS) and the second circumferentially asymmetric stiffness (CAS). **Figure 3** shows a rectangular cross-sectional beam segment with both ***CUS*** and ***CAS*** layup configurations. CUS layup configuration is manufactured by warping the composite layup using filament winding technique, $\theta(-z) = \theta(z)$, while CAS layup configuration is manufactured such that the beam cross section is symmetric about the *OXY* plane. $\theta(-z) = -\theta(z)$.

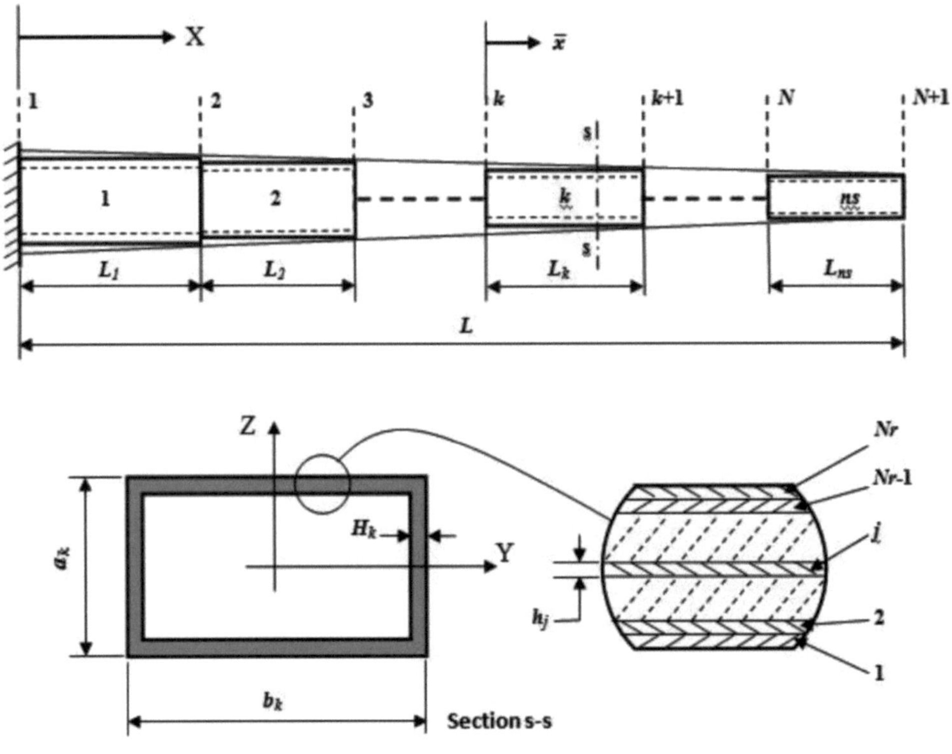

Figure 2.
Blade spar structural model.

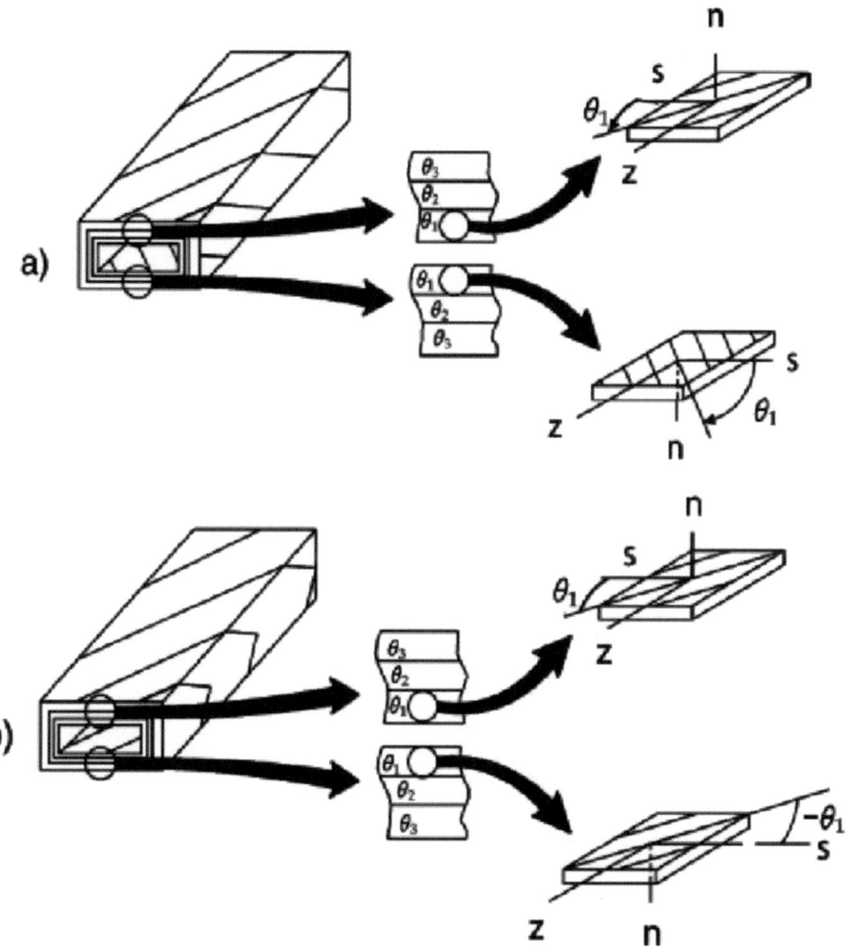

Figure 3.
Spar segment with (a) CUS and (b) CAS layup configurations.

2.1 CUS layup configuration

For any segment $\boldsymbol{k}$ within the blade structure, a local coordinate system with $\acute{X}$-axis is introduced such that (see **Figure 2**).

$$0 \le \acute{x} = X - X_k \le L_k \tag{3}$$

In the special case of CUS layup, the equations of motion reduce to (for the $\boldsymbol{k}^{th}$ segment):

$$\begin{aligned} & C_{11}^k U_1'' + C_{12}^k \varphi'' - m_k \acute{U}_1 = 0 \\ & C_{12}^k U_1'' + C_{22}^k \varphi'' - I_k \acute{\varphi} = 0 \\ & C_{33}^k U_3'''' + m_k \acute{U}_3 = 0 \\ & C_{44}^k U_2'''' + m_k \acute{U}_2 = 0 \end{aligned} \tag{4}$$

The first two equations express a coupled extension-twist vibration (ETV) mode, while the third and fourth equations express vertical bending vibration (VBV) and horizontal bending vibration (HBV) modes, respectively.

Applying the integral formulas given in Ref. [15], the stiffness coefficients can be determined from:

$$\begin{aligned}
C_{11}^{k} &= 2K_{A}^{k}(b_k + a_k - 2H_k) \\
C_{12}^{k} &= K_{B}^{k}(b_k - H_k)(a_k - H_k) \\
C_{22}^{k} &= K_{C}^{k}\frac{((b_k - H_k)(a_k - H_k))^2}{2(b_k + a_k - 2H_k)} \\
C_{33}^{k} &= \left(K_{A}^{k} - \frac{{K_{B}^{k}}^{2}}{K_{C}^{k}}\right)\left(\frac{(a_k - H_k)^3}{6}\right)\left(1 + \frac{3(b_k - H_k)}{(a_k - H_k)}\right) \\
C_{44}^{k} &= \left(K_{A}^{k} - \frac{{K_{B}^{k}}^{2}}{K_{C}^{k}}\right)\left(\frac{(b_k - H_k)^3}{6}\right)\left(1 + \frac{3(a_k - H_k)}{(b_k - H_k)}\right)
\end{aligned} \tag{5}$$

where K_A, K_B, and K_C are the reduced axial, coupled axial-shear, and shear stiffness coefficients, respectively, given by the expressions:

$$\begin{aligned}
K_A &= A_{11} - \frac{(A_{12})^2}{A_{22}} \\
K_B &= 2\left[A_{16} - \frac{A_{12}A_{26}}{A_{22}}\right] \\
K_C &= 4\left[A_{66} - \frac{(A_{26})^2}{A_{22}}\right]
\end{aligned} \tag{6}$$

where A_{mn} are called the membrane in-plane stiffnesses, which depend on the fiber orientation angle θ, volume fractions, and mechanical properties of the fiber and matrix materials [4].

The Young's moduli in the longitudinal and lateral directions of the lamina E_{11} and E_{22}, the shear modulus G_{12}, and the major Poisson's ratio ν_{12} are calculated using the semiempirical methods by Halpin and Tsai [23]:

$$\begin{aligned}
E_{11} &= E_m(1 - V_f) + E_f V_f \\
E_{22} &= E_m(1 + \eta_E V_f)/(1 - \eta_E V_f) \\
G_{12} &= G_m(1 + \eta_G V_f)/(1 - \eta_G V_f) \\
\nu_{12} &= \nu_m(1 - V_f) + \nu_f V_f \\
\eta_E &= (E_f - E_m)/(E_f + E_m) \\
\eta_G &= (G_f - G_m)/(G_f + G_m)
\end{aligned} \tag{7}$$

Subscripts "m" and "f" refer to the properties of matrix and fiber materials, respectively, and V_f is the volume fraction of fibers within each lamina. Considering the equations of coupled extension-twist vibration, the assumed solution is [24]:

$$\begin{aligned}
U_1(x,t) &= \acute{C}_1 e^{\lambda x + i\omega t} \\
\varphi(x,t) &= \acute{C}_2 e^{\lambda x + i\omega t}
\end{aligned} \tag{8}$$

where ω is the circular natural frequency of free vibration. Substituting from Eq. (8) into Eq. (4), the associated characteristic equation can be shown to be:

$$\begin{aligned} &a\lambda^4 + b\omega^2\lambda^2 + c\omega^4 = 0 \\ &a = C_{11}^k C_{22}^k - {C_{12}^k}^2 \\ &b = C_{11}^k I_k + C_{22}^k m_k \\ &c = m_k I_k \end{aligned} \tag{9}$$

The solution can be expressed as:

$$\begin{aligned} &\lambda_{1,2} = \pm i\alpha_1 \\ &\lambda_{3,4} = \pm i\alpha_2 \\ &\alpha_{1,2} = \omega\sqrt{\frac{b \mp \sqrt{b^2 - 4ac}}{2a}} \end{aligned} \tag{10}$$

If $C_{11}^k I_k > C_{22}^k m_k$, then the minus sign in Eq. (10) will generate natural frequencies with axial-mode dominated vibration, while the plus sign will generate natural frequencies with torsion-mode dominated vibration. If $C_{11}^k I_k < C_{22}^k m_k$, the domination is reversed. The general solution of the coupled extension-twist vibration problem (Eq. 4) can be expressed as:

$$\begin{aligned} &U_1(x,t) = (c_1 \sin\alpha_1 x + c_2 \cos\alpha_1 x + c_3 \sin\alpha_2 x + c_4 \cos\alpha_2 x)e^{i\omega t} \\ &\varphi(x,t) = (c_5 \sin\alpha_1 x + c_6 \cos\alpha_1 x + c_7 \sin\alpha_2 x + c_8 \cos\alpha_2 x)e^{i\omega t} \end{aligned} \tag{11}$$

In order to satisfy both the coupled equations of motion, there are specific relations between the constants such that:

$$\begin{aligned} &c_5 = q_1 c_1, c_6 = q_1 c_2, c_7 = q_2 c_3, c_8 = q_2 c_4 \\ &q_{1,2} = \frac{2m_k a}{C_{12}^k\left(b \mp \sqrt{b^2 - 4ac}\right)} - \frac{C_{11}^k}{C_{12}^k} \end{aligned} \tag{12}$$

2.2 CAS layup configuration

In this case, the equations of motion, applied to segment $\boldsymbol{k}$, reduced to:

$$\begin{aligned} &C_{11}^k U_1'' - m_k \acute{U}_1 = 0 \\ &C_{22}^k \varphi'' + C_{23}^k U_3''' - I_k \acute{\varphi} = 0 \\ &C_{23}^k \varphi''' + C_{33}^k U_3'''' + m_k \acute{U}_3 = 0 \\ &C_{44}^k U_2'''' + m_k \acute{U}_2 = 0 \end{aligned} \tag{13}$$

The second and third equations of motion express a coupled bending-twist vibration (BTV) mode, while the first and fourth equations of motion express extension vibration (EV) and horizontal bending vibration (HBV) modes, respectively. The non-zero stiffness coefficients are given by:

$$
\begin{aligned}
C_{11}^{k} &= 2K_{A}^{k}(b_k + a_k - 2H_k) - 2\frac{K_{B}^{k\,2}}{K_{C}^{k}}(b_k - H_k) \\
C_{22}^{k} &= K_{C}^{k}\frac{((b_k - H_k)(a_k - H_k))^2}{2} \\
C_{23}^{k} &= K_{B}^{k}\frac{((b_k - H_k)(a_k - H_k))^2}{2} \\
C_{33}^{k} &= K_{A}^{k}\frac{(a_k - H_k)^3}{6}\left[1 + \frac{3(b_k - H_k)}{(a_k - H_k)}\right] - \frac{K_{B}^{k\,2}}{2K_{C}^{k}} \\
C_{44}^{k} &= \frac{(b_k - H_k)^3}{6}\left[\left(1 + \frac{3(a_k - H_k)}{(b_k - H_k)}\right)K_{A}^{k} - \frac{K_{B}^{k\,2}}{K_{C}^{k}}\right]
\end{aligned}
\tag{14}
$$

Assuming harmonic solution similar to that given in Eq. (8), the characteristic equation is obtained by differentiating Eq. (13) to get:

$$
\begin{aligned}
&a\lambda^6 + b\omega^2\lambda^4 - c\omega^2\lambda^2 - d\omega^4 = 0 \\
&a = C_{22}^{k}C_{33}^{k} - C_{23}^{k\,2} \\
&b = C_{33}^{k}I_k \\
&c = C_{22}^{k}m_k \\
&d = m_k I_k
\end{aligned}
\tag{15}
$$

which has the solution:

$$
\lambda_{1,2} = \pm i\gamma_1, \lambda_{3,4} = \pm i\gamma_2, \lambda_{5,6} = \pm\gamma_3 \tag{16}
$$

The general solution takes the form:

$$
\begin{aligned}
U_3(x,t) &= (c_1 \sin\gamma_1 x + c_2 \cos\gamma_1 x + c_3 \sin\gamma_2 x + c_4 \cos\gamma_2 x + c_5 \sinh\gamma_3 x + c_6 \cosh\gamma_3 x)e^{i\omega t} \\
\varphi(x,t) &= (c_7 \sin\gamma_1 x + c_8 \cos\gamma_1 x + c_9 \sin\gamma_2 x + c_{10} \cos\gamma_2 x + c_{11} \sinh\gamma_3 x + c_{12} \cosh\gamma_3 x)e^{i\omega t}
\end{aligned}
\tag{17}
$$

The relations between the constants are given by:

$$
\begin{aligned}
&c_7 = k_1 c_2, c_9 = k_2 c_4, c_{11} = k_3 c_6 \\
&c_8 = -k_1 c_1, c_{10} = -k_2 c_3, c_{12} = k_3 c_5 \\
&k_1 = \frac{C_{23}^{k}\gamma_1^3}{C_{22}^{k}\gamma_1^2 - I_k\omega^2} \\
&k_2 = \frac{C_{23}^{k}\gamma_2^3}{C_{22}^{k}\gamma_2^2 - I_k\omega^2} \\
&k_3 = \frac{-C_{23}^{k}\gamma_3^3}{C_{22}^{k}\gamma_3^2 + I_k\omega^2}
\end{aligned}
\tag{18}
$$

3. Natural frequencies and mode shapes

Applying the transfer-matrix method [25], the relation between the state vectors at one end of a segment $\{Z\}_k$ and that at the other end $\{Z\}_{k+1}$ is given by:

$$\{Z\}_{k+1} = [T]_k \{Z\}_k \tag{19}$$

where $[T]_k$ is called the elementary transfer matrix associated with the kth segment. For a spar built up of *Ns* uniform segments, Eq. (19) can be applied at successive stations to obtain:

$$\begin{aligned} \{Z\}_{Ns+1} &= [T]_o \{Z\}_1 \\ [T]_o &= [T]_{Ns}[T]_{Ns-1} \cdots \cdots \cdots [T]_k \cdots \cdots \cdots [T]_2[T]_1 \end{aligned} \tag{20}$$

where the matrix $[T]_o$ is known as the overall transfer matrix, which relates the state vectors at the spar fixed end to the free end at which the boundary conditions are specified. Therefore, applying the boundary conditions at both ends and considering only the nontrivial solutions, the frequency equations can be readily obtained. Derivations of the elementary transfer matrices for different vibration modes are discussed in the following section.

3.1 CUS layup

In this case the elements of the state vector at both ends of the kth segment are related by the elementary transfer matrix for the coupled extension-twist vibration as follows:

$$\begin{bmatrix} U_1(L_k) \\ \varphi(L_k) \\ C_{11}U_1'(L_k) + C_{12}\varphi'(L_k) \\ C_{22}\varphi'(L_k) + C_{12}U_1'(L_k) \end{bmatrix} = \begin{bmatrix} T_{11} & T_{12} & T_{13} & T_{14} \\ T_{21} & T_{22} & T_{23} & T_{24} \\ T_{31} & T_{32} & T_{33} & T_{34} \\ T_{41} & T_{42} & T_{43} & T_{44} \end{bmatrix}_k \begin{bmatrix} U_1(0) \\ \varphi(0) \\ C_{11}U_1'(0) + C_{12}\varphi'(0) \\ C_{22}\varphi'(0) + C_{12}U_1'(0) \end{bmatrix} \tag{21}$$

Imposing the cantilevered boundary conditions on the overall transfer matrix, the overall frequency equation for coupled extension-twist vibration of the blade spar is given by:

$$T^o_{33}T^o_{44} - T^o_{34}T^o_{43} = 0 \tag{22}$$

3.2 CAS layup

Here the transfer matrix equation for the segment *k* is:

$$\begin{bmatrix} -U_3(L_k) \\ \varphi(L_k) \\ U_3'(L_k) \\ C^k_{22}\varphi'(L_k) + C^k_{23}U_3''(L_k) \\ C^k_{23}\varphi'(L_k) + C^k_{33}U_3''(L_k) \\ C^k_{33}U_3'''(L_k) \end{bmatrix} = \begin{bmatrix} T_{11} & T_{12} & T_{13} & T_{14} & T_{15} & T_{16} \\ T_{21} & T_{22} & T_{23} & T_{24} & T_{25} & T_{26} \\ T_{31} & T_{32} & T_{33} & T_{34} & T_{35} & T_{36} \\ T_{41} & T_{42} & T_{43} & T_{44} & T_{45} & T_{46} \\ T_{51} & T_{52} & T_{53} & T_{54} & T_{55} & T_{56} \\ T_{61} & T_{62} & T_{63} & T_{64} & T_{65} & T_{66} \end{bmatrix}_k \begin{bmatrix} -U_3(0) \\ \varphi(0) \\ U_3'(0) \\ C^k_{22}\varphi'(0) + C^k_{23}U_3''(0) \\ C^k_{23}\varphi'(0) + C^k_{33}U_3''(0) \\ C^k_{33}U_3'''(0) \end{bmatrix} \tag{23}$$

Imposing the cantilevered boundary conditions on the overall transfer matrix, the frequency determinant is given by:

$$\begin{vmatrix} T_{44} & T_{45} & T_{46} \\ T_{54} & T_{55} & T_{56} \\ T_{64} & T_{65} & T_{66} \end{vmatrix}_o = 0 \tag{24}$$

The natural frequencies for coupled bending-twist vibration can be obtained numerically by solving the characteristic determinant of Eq. (24) for γ.

4. Optimization model formulation

4.1 Selection of design variables and pre-assigned parameters

In order to formulate a practical and inexpensive optimization model, the large number of design variables of a wind turbine blade has to be reduced to a reasonable number that can be easily dealt with. The design variables which are not subject to change in the optimization process are called the pre-assigned parameters. They are selected to be:

a. Type of materials of construction

b. Total blade length

c. Chord distribution along the blade axis

d. Twist angle distribution along the blade axis

e. Airfoil type and dimensions

f. Dimensions and spacing of the internal supporting ribs

g. Covering skin thickness

h. Blade-to-hub attachment, which is chosen to be of hingeless type

i. Shape of the spar cross section, which is chosen to be rectangular

On the other hand, the design variables, which are subject to change during the optimization process, are chosen to be the fiber orientation angle $\theta_{j,k}$, fiber volume fraction $V_{fj,k}$, thickness of each lamina $h_{j,k}$, cross-sectional dimensions (a_k, b_k, H_k), and length L_k of each segment composing the blade spar (refer to **Figure 2**). The cross-sectional aspect ratio (a_k/b_k) can be prescribed according to the airfoil cross-sectional dimensions, and only the variable a_k is taken as a design variable. In addition, to formulate a normalized optimization model with scaled variables, it is necessary to start with a known reference beam to which all the design variables, constraints, and objective function are referred. It is chosen to be made of unidirectional laminated composites with fiber volume fraction V_{fo} and fiber orientation angle θ_o, and its cross-sectional dimensions are denoted by b_o for width, a_o for depth, H_o for wall thickness, and L_o for length. Therefore, the normalized variables are defined as follows:

$$\begin{aligned} \hat{L}_k &= L_k/L_o \\ \hat{b}_k &= b_k/b_o \\ \hat{a}_k &= a_k/a_o \\ \hat{H}_k &= H_k/H_o = \sum_{j=1}^{Nr} \hat{h}_j \\ \hat{h}_j &= h_j/H_o \end{aligned} \tag{25}$$

For a spar beam composed of *Ns* segments, the design variables are defined by the following matrix equation:

$$[X] = \begin{bmatrix} \hat{L}_1 & \hat{L}_2 & & \hat{L}_k & & \hat{L}_{Ns} \\ \hat{a}_1\theta_{j,1} & \hat{a}_2\theta_{j,2} & & \hat{a}_k\theta_{j,k} & & \hat{a}_{Ns}\theta_{j,Ns} \\ V_{f\,j,1} & V_{f\,j,2} & & V_{f\,j,k} & & V_{f\,j,Ns} \\ \hat{h}_{j,1} & \hat{h}_{j,2} & & \hat{h}_{j,k} & & \hat{h}_{j,Ns} \\ \vdots & \vdots & \cdots\cdots & \vdots & \cdots\cdots & \vdots \\ \vdots & \vdots & & \vdots & & \vdots \\ \vdots & \vdots & & \vdots & & \vdots \\ \theta_{Nr,1} & \theta_{Nr,2} & & \theta_{Nr,k} & & \theta_{Nr,Ns} \\ V_{f\,Nr,1} & V_{f\,Nr,2} & & V_{f\,Nr,k} & & V_{f\,Nr,Ns} \\ \hat{h}_{Nr,1} & \hat{h}_{Nr,2} & & \hat{h}_{Nr,k} & & \hat{h}_{Nr,Ns} \end{bmatrix} \tag{26}$$

The total number of design variables in this case is equal to (3 Nr + 2)*Ns, which is a relatively large number of optimization variables.

4.2 Different optimization strategies

Several criteria are used in the dynamic optimization of wind turbine blades; some important ones are the minimal mass design, maximum frequency design, and frequency-placement criteria. In the first strategy, the mass of the blade is minimized while imposing constraints on the blade natural frequencies and strength. In the maximum frequency criterion, the reduction of vibration level is attained by maximizing the natural frequencies of the blade without regard to the complicated stiffness/mass ratio and the exciting frequencies constraints. Higher natural frequencies are favorable for reducing both of the steady-state and transient responses of any structure being excited. The last category of the objective functions to be considered is the placement of the blade natural frequencies, where the main goal is to separate the natural frequencies of the blade from the exciting frequencies to avoid large amplitudes from occurring near the resonant conditions.

4.2.1 Minimal mass design

The minimal mass design optimization problem can be stated as follows:

$$\begin{array}{ll} \text{Find} & [X] \\ \text{Which} & \text{minimizes} f(X) = \hat{m} \\ \text{Subject to:} & \\ \text{Strength constraint} & \alpha_T - 1 \le 0 \\ \text{Resonance avoidance} & (1 - \Delta_n) \le \dfrac{\hat{\omega}_n}{\hat{\hat{\omega}}_n} \le (1 + \Delta_n), n = 1, 2, \ldots \\ \text{Side constraints} & [X_L] \le [X] \le [X_U] \end{array} \tag{27}$$

where $\hat{m} = m/m_o$ and $\hat{\omega}_n = \omega_n/\omega_{no}$ are the dimensionless mass and frequency and m_o and ω_{no} are the mass and frequency of the reference beam, respectively. $[X_L] \wedge [X_U]$ are the lower and upper bounds imposed on the design variables.

$$\begin{array}{l} \text{Structural mass}: m = 2\sum_{k=1}^{ns}\left[\left(\sum_{j=1}^{nr} \rho_j h_j\right)(b_k + a_k - 2H_k)L_k\right] \\ m_o = 2\rho_o H_o (b_o + a_o - 2H_o)L_o \end{array} \tag{28}$$

The mass densities ρ_o and ρ_j are calculated according to the volume fractions V_{fo} and V_{fj}, respectively. The symbolα_T is called the Tsai-Hill rupture coefficient [26], which depends on the ratios between the principle and rupture stresses (more details are given in appendix A). $\hat{\hat{\omega}}_n$, n = 1,2, ... are called the target (or desired) frequencies of a known baseline design adjusted to be well separated from the exciting frequencies, and the increments $\Delta_n$are the associated allowable tolerance of each frequency (e.g., Δ_n = 1%).

4.2.2 Maximum frequency optimization

Minimization of the overall vibration level is one of the most cost-effective solutions for a successful wind turbine design. It enhances other important design goals such as long fatigue life, high stability, and low noise level. Reduction of the overall vibration level can be attained by maximizing the stiffness-to-weight ratio of the wind turbine blade spar. It is well-known that natural frequency is a good indicator of structural stiffness-to-mass ratio. Thus, the optimization problem considered in this investigation will seek maximization of natural frequencies of the blade spar for different modes of vibrations while maintaining its total structural mass lower than or equal to that of a baseline design. Constraints are imposed on the optimization problem such that the blade spar has to be enclosed by the outer blade skin and has the sufficient strength to carry the applied loads without mass penalty.

The maximum frequency criterion, taking into considerations the different modes of vibration, may be cast in the following form:

$$\begin{array}{ll} \text{Find} & [X] \\ \text{Which} & \text{minimizes} f(X) = -\sum_n W_n \hat{\omega}_n, n = 1, 2, \ldots \\ \text{Subject to :} & \\ \text{Strength constraint} & \alpha_T - 1 \le 0 \\ \text{Mass constraint} & \hat{m} - 1 \le 0 \\ \text{Side constraints} & [X_L] \le [X] \le [X_U] \end{array} \tag{29}$$

The function $f(X)$ represents a weighted sum of the nondimensional natural frequencies of successive modes of vibration with $W_n$the weighting factor of the nth frequency $\hat{\omega}_n$.

4.2.3 Frequency placement

A good design philosophy for vibration reduction is to separate the natural frequencies of the structure from the harmonics of other excitation sources. Thus, natural frequencies of the blade spar should be placed near a target value. Mathematically, the frequency-placement criterion may be expressed by minimizing an objective function constructed from a weighted sum of the squares of the differences between each important frequency $\hat{\omega}_n$ and its target value $\hat{\omega}_n$ as follows:

$$\begin{array}{ll} \text{Find} & [X] \\ \text{Which} & \text{minimizes} f(X) = \sum_n W_n (\hat{\omega}_n - \hat{\omega}_n)^2 \\ \text{Subject to:} & \\ \text{Strength constraint} & \alpha_T - 1 \leq 0 \\ \text{Mass constraint} & \hat{m} - 1 \leq 0 \\ \text{Side constraints} & [X_L] \leq [X] \leq [X_U] \end{array} \tag{30}$$

4.3 Side constraints

Side constraints are imposed on the design variables in order to verify various geometric, manufacturing, or logical reasons. In the present optimization model, these constraints are defined in the following.

4.3.1 Length of blade spar

The total length of the blade is kept equal to that of a known baseline design:

$$\left(\sum_{k=1}^{Ns} \hat{L}_k\right) - 1 = 0 \qquad \hat{L}_k \geq 0 \tag{31}$$

4.3.2 Spar cross-sectional dimensions

Since the blade spar is limited by a tapered configuration with a certain type of airfoil cross sections, the height of any segment k having its end at a specific position x_{k+1} (refer to **Figure 2**) must not exceed the outermost height (a_U) at this position. **Figure 4** shows the outermost dimensions of a wind turbine blade spar of length L, where a_r and a_t represent the spar heights at the root and tip locations, respectively. The dimensionless upper limiting height $\hat{a}_U$ is defined by the relation:

$$\hat{a}_U = \hat{a}_r\left(1 - \left(1 - \frac{a_t}{a_r}\right)\hat{x}_{k+1}\right) \qquad \hat{x}_{k+1} = \hat{L}_1 + \hat{L}_2 + \ldots + \hat{L}_k \tag{32}$$

Lower bounds are also imposed on the height of the kth segment as a reasonable percentage of the airfoil height near the tip ($\hat{a}_L = 0.5\hat{a}_t$). Thus, the inequality

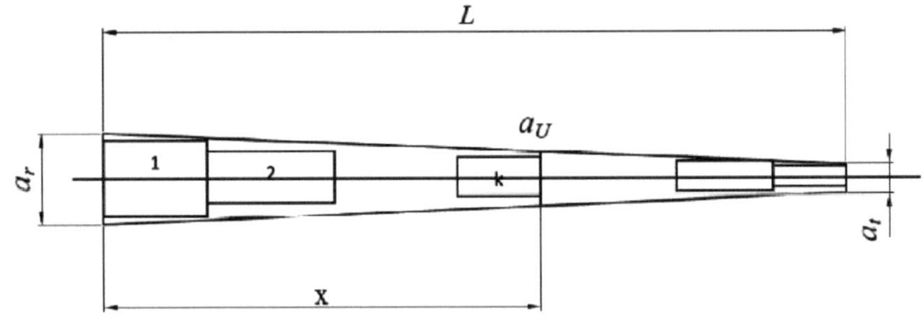

Figure 4.
Outermost dimensions of a wind turbine blade spar.

constraints imposed on the dimensionless height of each segment of the blade spar are expressed as:

$$\hat{a}_L \leq \hat{a}_k \leq \hat{a}_U, k = 1, 2, \ldots .Ns \tag{33}$$

Upper bounds on the segment height represent the main obstacle in blade spar optimization since it restricts the spanwise variation in spar height. Note that the width of each segment (b_k) depends on the height of that segment (a_k) according to the pre-assigned aspect ratio of the cross section.

In addition, the inequality constraints imposed on the walls thickness of each segment are:

$$\hat{H}_L \leq \hat{H}_k \leq \hat{H}_U, k = 1, 2, \ldots .Ns \tag{34}$$

4.3.3 Fiber volume fraction and orientation angle within any segment (k)

$$(0.2, -\pi/2) \leq \left(V_{fj}, \theta_j\right) \leq (0.8, \pi/2), j = 1, 2, \ldots Nr \tag{35}$$

The optimization models described by Eqs. (27–35) belong to the constrained nonlinear mathematical programming models since the objective functions and most of the constraints contain nonlinear algebraic expressions. Such problems can be solved by a variety of mathematical programming techniques [27, 28]. The sequential quadratic programming (SQP) is one of the most powerful developed and perhaps one of the best methods of optimization. The method has a theoretical basis that is related to the solution of a set of nonlinear equations using Newton's method and the derivation of simultaneous nonlinear equations using Karush-Kuhn-Tucker (KKT) conditions to the Lagrangian of the constrained optimization problem [29, 30]. More details are given in Appendix B.

5. Behavior of the objective function

Focusing on CUS layup configurations, extensive studies have been carried out, using MATLAB, on the objective functions in order to be able to visualize the unconstrained behavior of natural frequencies and mass of specific thin-walled composite beams. Studies are performed in a two-dimensional design space such that only two design variables are allowed to change while the others are assigned to specific values. All variables are normalized with respect to a known reference beam parameters, as given in Eq. (25).

5.1 Case of two-segment cantilever beam

For a two-segment beam with each segment constructed only from one layer, the matrix of variables (Eq. 26) can be reduced to take the form:

$$[X] = \begin{bmatrix} \hat{L}_1 & \hat{L}_2 \\ \hat{a}_1 & \hat{a}_2 \\ \theta_{1,1} & \theta_{1,2} \\ V_{f_{1,1}} & V_{f_{1,2}} \\ \hat{h}_{1,1} & \hat{h}_{1,2} \end{bmatrix} \tag{36}$$

Note that in this case $\hat{H}_k = \hat{h}_j$. The effects of changing two corresponding variables simultaneously in the two segments are studied in the following.

5.1.1 *Effect of changing fiber orientation angles within the two segments (θ_1, θ_2)*

Consider a two-segment beam with a height of the outboard segment half that of the inboard segment but with equal length. Fiber volume fraction and wall thickness of the two segments are the same as that of the reference beam. The following matrix of pre-assigned variables describes this case:

$$[X] = \begin{bmatrix} 0.5 & 0.5 \\ 1.5 & 0.75 \\ \theta_1 & \theta_2 \\ 0.5 & 0.5 \\ 1 & 1 \end{bmatrix} \tag{37}$$

Figure 5 shows the developed contours of the normalized fundamental bending frequency in the (θ_1, θ_2)-design space. It is noticed that the maximum bending natural frequencies occurs at the point $(\theta_1, \theta_2) = (0, 0)$, which is an expected natural solution. **Figure 6** shows the level curves of the normalized twist-dominated natural frequency. It indicates that the maximum occurs near the design point $(\theta_1, \theta_2) = (45, 45)$. On the other hand, the design points (θ_1, θ_2) = (0, 0), (90, 90), (0, 90), and (90, 0) give the same values of the frequency representing a local minima solution, as shown in **Figure 6**. The angle sequence in this case affects the frequency level curves due to the change in heights of the two segments.

5.1.2 *Effect of changing length and height of the outboard segments ($\hat{L}_2$, $\hat{a}_2$)*

Consider a two-segment beam with the following matrix of pre-assigned variables. The height and thickness of the inboard segment are equal to those of the reference beam, while thickness of the outboard segment is half that of the reference beam. Nondimensional length of the inboard segment is $1 - \hat{L}_2$ in order to keep the total length equal to that of the reference beam:

$$[X] = \begin{bmatrix} 1-\hat{L}_2 & \hat{L}_2 \\ 1 & \hat{a}_2 \\ 0 & 0 \\ 0.5 & 0.5 \\ 1 & 0.5 \end{bmatrix} \tag{38}$$

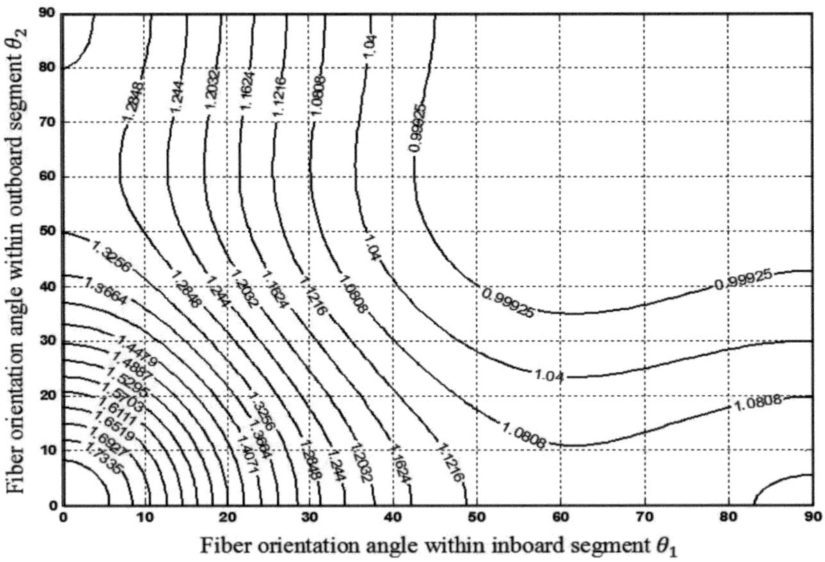

Figure 5.
Effect of fiber orientation angles on the normalized fundamental bending frequency (a two-segment cantilevered beam).

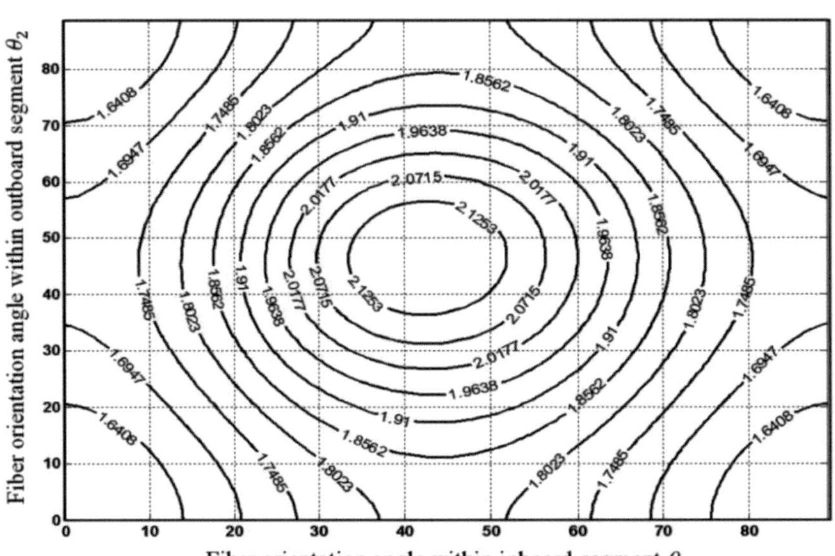

Figure 6.
Effect of fiber orientation angles on normalized twist-dominated natural frequency (a two-segment cantilevered beam).

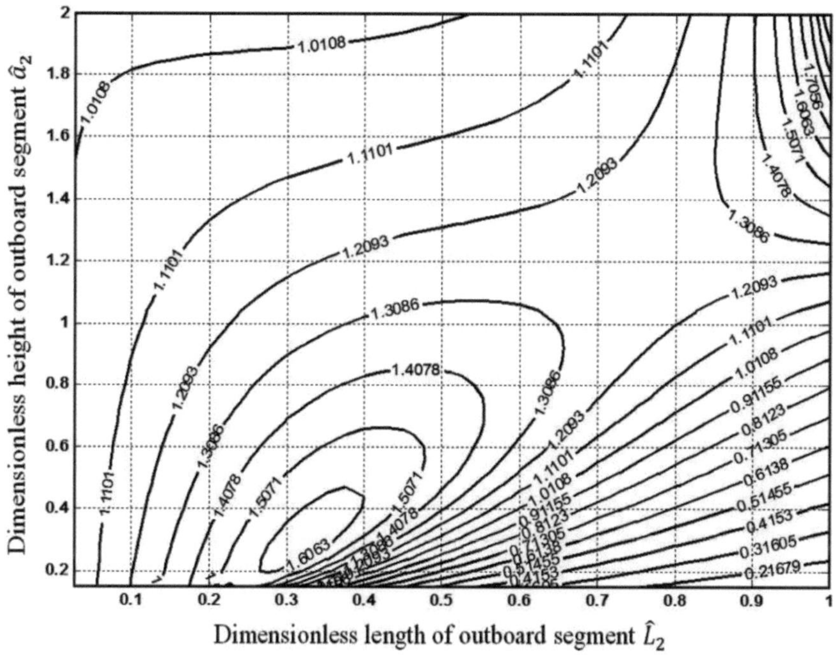

Figure 7.
Effect of length and height of the outboard segment on normalized fundamental bending frequency (a two-segment cantilevered beam).

Figure 7 shows the developed contours of the fundamental bending frequency in the $(\hat{L}_2, \hat{a}_2)$-design space. It is seen that the frequency domain is divided into three regions bonded by the reference frequency value. The extreme left and right regions contain two local minima. The central region contains the global maximum. The maximum bending natural frequencies occur near the point $(\hat{L}_2, \hat{a}_2) \approx (0.35, 0.3)$. This point corresponds to a total structural mass of about 70% of the reference beam, which represents a significant gain from the optimization process. This configuration of beam dimensions gives a tapered beam with a decreasing height and length. The same value of maximum natural frequency is also repeated in the design space at the uppermost right region at $(\hat{L}_2, \hat{a}_2) \approx (1, 2)$, but the total mass becomes higher and equal to that of the reference beam.

From the previous investigations, it is observed that the frequency functions, even though implicit in the design variables, are well behaved and continuous in the selected design spaces. A weighted sum of successive natural frequencies is a good representative for the objective function of the optimization model. Another important observation is that the length and the height of the individual segments are so effective variables in the whole optimization process.

6. Case study

As a case study, the main spar of a medium scale composite blade of a 750-kW horizontal axis wind turbine is optimized. Full description and technical data can be found in Ref. [31]. The blades cross section having NACA 63-218 airfoil is made of E-glass/epoxy composites with properties given in **Table 1**. The effective length of the spar is 18.33 m extending from 20–98% of the blade length. The spar cross-sectional dimensions at root are 340 mm height and 680 mm width and at the tip 100 mm and 200 mm. The wall thickness at the root and tips are 15 mm and 5 mm, respectively. The chordwise location of the spar cross-sectional center coincides with the blade pitch axis.

6.1 Baseline design

An initial baseline design of the wind turbine blade spar with total mass = 580.55 kg is selected to be composed of 10 segments of equal length. Fiber orientation angle and volume fraction within each segment are pre-assigned to be *0°* and *50%*, respectively. Structural and dynamic properties of the spar baseline design are given in **Tables 2** and **3**, respectively.

Property	Epoxy matrix	E-glass fibers
Modulus of elasticity (GPa)	$E_m = 4.5$	$E_{11f} = E_{22f} = 74$
Modulus of rigidity (GPa)	$G_m = 1.6$	$G_{12f} = 30$
Poisson's ratio	$\nu_m = 0.4$	$\nu_{12f} = 0.25$
Density (kg/m^3)	$\rho_m = 1200$	$\rho_f = 2600$
Tensile strength (MPa)	$\sigma_{mr} = 90$	$\sigma_{11r}, \sigma_{22r}$ (Eq. A-2)
Shear strength (MPa)	$\tau_{mr} = 52$	

Table 1.
Material properties of the spar structure [4, 31].

Segment no.	Length (mm)	Height (mm)	Width (mm)	Thickness (mm)
1	1833	316	632	14.4
2		292	584	13.3
3		268	536	12.2
4		244	488	11.1
5		220	440	10
6		196	392	8.9
7		172	344	7.8
8		148	296	6.7
9		124	248	5.6
10		100	200	4.5

Table 2.
Dimensions of the baseline design.

Vibration mode (dominated)	Natural frequencies (Hz)		
	First mode	Second mode	Third mode
Twist	36.73	64.88	95.18
Extension	94.9	203.5	319.8
Flap bending	1.55	4.90	11.15

Table 3.
Natural frequencies of the baseline design.

The calculated normal and tangential distributed forces (P_z, P_y) and the corresponding bending moments (M_z, M_y) applying to the wind turbine blade are shown in **Figures 8** and **9**, respectively (refer to Appendix C). The wind speed of 60 m/s (survival loading condition) is taken in calculating the applied loads, while the blades were considered in the stationary horizontal position. These forces and moments are implemented in the strength constraints of the present optimization model.

6.2 Definition of the reference beam

The reference beam, to which all the design variables and frequencies are normalized, is selected to be a uniform, thin-walled cantilevered spar with rectangular

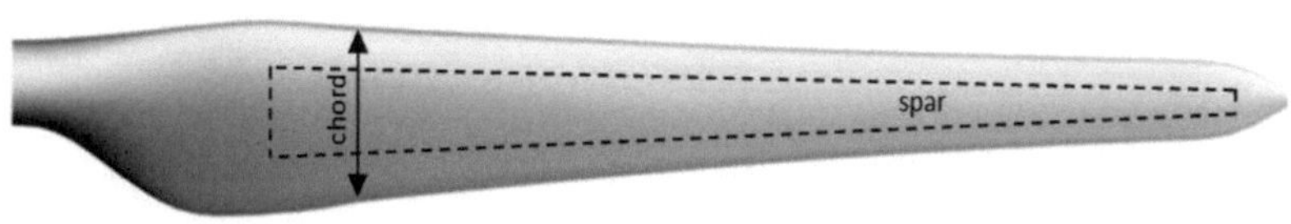

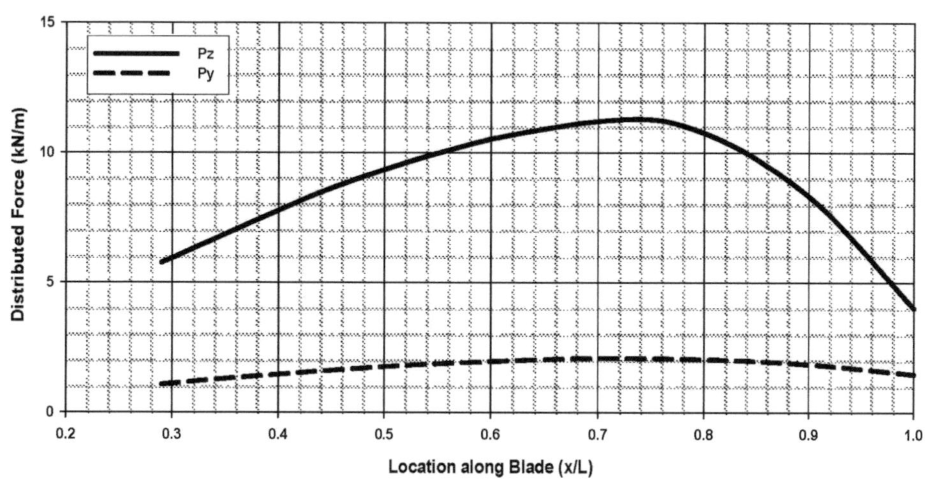

Figure 8.
Shear forces applied to the blade at 60-m/s wind speed.

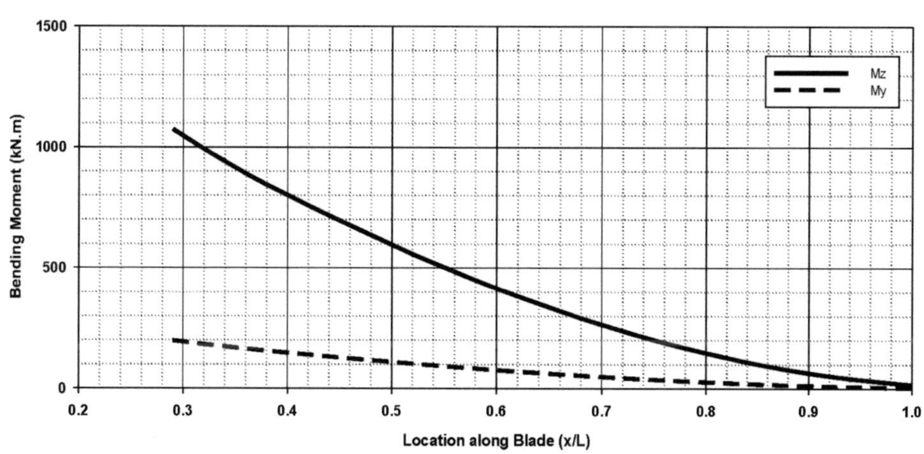

Figure 9.
Flapwise and chordwise bending moments applied to the blade at 60-m/s wind speed.

Dimensions (mm)	Natural frequencies: $\omega_{n_o}/2$ (Hz)			
	Mode	First	Second	Third
Width b_o = 440				
Height a_o = 220	Twist	17.59	52.76	87.93
Length L_o = 18,330	Extension	61.99	185.97	309.95
Thickness H_o = 10	Bending	0.65	4.07	11.39

Table 4.
Dimensions and natural frequencies of the reference beam.

Design variable	Lower boundary	Upper boundary	Starting value
Dimensionless length of each segment $\hat{L}_k$	0.0	1	1/*Ns*
Dimensionless height of each segment $\hat{a}_k$	0.2	2	1
Fiber orientation angle $\theta_{j,k}$	$-\pi/2$	$\pi/2$	0
Fiber volume fraction $V_{f\,j,k}$	0.2	0.8	0.5
Dimensionless thickness of each layer $\hat{h}_{j,k}$	0.01	0.2	0.1

Table 5.
Lower, upper, and starting boundaries of design variables for multiple-segment spar.

cross section constructed from single unidirectional E-glass/epoxy composite layer (θ_o = 0) and fiber volume fraction of V_{fo} = 50%. This fiber layup configuration gives maximum natural frequencies for bending and extension modes of vibration while maintaining moderate structural mass (m_o = 580.5 *kg*) and strength. The dimensions of the reference beam are selected according to the mean cross-sectional dimensions of the spar baseline design (see **Table 4**).

6.3 Optimization results

The blade spar described in Section 6.1 is optimized according to the three different optimization models under the specified constrains as given in Section 4.2. The method of sequential quadratic programming with MATLAB optimization toolbox is applied to obtain the needed optimal solutions (refer to Appendix B). The starting, lower, and upper bounds for each design variable are given in **Table 5**. Determination of these values affects the optimization results; thus, several trials were performed in order to select the appropriate values.

6.3.1 Minimum mass design

In this formulation the target frequencies are set equal to the first three flapping frequencies of the baseline design, which are well separated from resonant frequencies. The attained optimal solutions are given in **Table 6** with the spar cross sections constructed from balanced and symmetric laminates. The achieved mass saving reached about 25%, where the natural frequencies are found to be too close to those of the baseline design, as given in **Table 7**. Considerable mass reduction can be observed with much savings for the inboard segments as compared to that for the outboard segments.

Segment no.	Segment length (mm)	Segment cross section (mm)		Fiber volume fraction	No. of layers	Layup
		Height	Thickness			
1	280.5	323.8	9.9	0.31	18	$(\pm45/0_3/\pm45/0_3/\pm45)_s$
2	183.3	316.7	9.9	0.33	18	$(\pm45/0_3/\pm45/0_3/\pm45)_s$
3	200.8	313.8	8.9	0.32	16	$(\pm45/0_3/\pm45/0_2/\pm45)_s$
4	534.1	301.4	8.9	0.33	16	$(\pm45/0_2/\pm45/0_3/\pm45)_s$
5	851.0	286.3	8.9	0.32	16	$(\pm45/0_3/\pm45/0_2/\pm45)_s$
6	1031.7	263.8	7.9	0.32	14	$(\pm45/0_2/\pm45/0_2/\pm45)_s$
7	1112.0	235.2	7.8	0.33	14	$(\pm45/0_2/\pm45/0_2/\pm45)_s$
8	1272.3	207.4	7.8	0.34	14	$(\pm45/0_5/\pm45)_s$
9	1482.5	188.9	6.8	0.33	12	$(\pm45/0_5)_s$
10	1589.4	170.4	6.5	0.33	12	$(\pm45/0_5)_s$
11	1273.1	154.0	6.5	0.33	12	$(\pm45/0_5)_s$
12	1132.9	138.6	6.4	0.32	12	$(\pm45/0_5)_s$
13	1100.9	120.9	5.7	0.33	10	$(\pm45/0_4)_s$
14	1051.8	106.3	5.6	0.33	10	$(\pm45/0_4)_s$
15	1177.6	94.1	5.2	0.33	8	$(\pm45/0_3)_s$
16	880.5	82.4	4.8	0.33	8	$(\pm45/0_3)_s$
17	693.2	70.7	4.6	0.33	8	$(\pm45/0_3)_s$
18	540.0	59.0	4.5	0.34	8	$(\pm45/0_3)_s$
19	456.8	48.0	4.0	0.33	6	$(\pm45/0_2)_s$
20	1485.7	44.0	3.6	0.20	6	$(\pm45/0_2)_s$

Table 6.
Minimum mass design optimization.

Vibration mode (dominated)	Design	Natural frequencies (Hz)			$\sum_{n=1}^{3} W_n\omega_n$
		First mode	Second mode	Third mode	
Twist	Baseline	36.73	64.88	94.90	50.49
	Min. mass	36.16	65.62	89.13	49.52
Extension	Baseline	94.90	203.50	319.80	148.00
	Min. mass	89.13	194.88	299.66	139.77
Flapping	Baseline	1.55	4.90	11.15	2.48
	Min. mass	1.53	4.88	11.17	2.45
Mass (kg)	Baseline		580.55		
	Min. mass		437.76		

Table 7.
Comparison of natural frequencies and mass between baseline and the minimum mass designs.

6.3.2 Maximum frequency design

In order to obtain a blade spar with higher natural frequencies, the maximum frequency optimization model given in Eq. 29 is applied under inequality mass

Segment no.	Segment length (mm)	Segment cross section (mm)		Fiber volume fraction	No. of layers	Layup
		Height	Thickness			
1	842	329	18.7	0.8	34	$(\pm45/0_7/\pm45/0_7/\pm45)_s$
2	873	318	16.3	0.8	30	$(\pm45/0_6/\pm45/0_6/\pm45)_s$
3	903	306	14.4	0.8	26	$(\pm45/0_5/\pm45/0_5/\pm45)_s$
4	919	294	12.3	0.8	22	$(\pm45/0_4/\pm45/0_4/\pm45)_s$
5	932	281	11	0.8	20	$(\pm45/0_3/\pm45/0_4/\pm45)_s$
6	937	269	8.8	0.8	16	$(\pm45/0_3/\pm45/0_2/\pm45)_s$
7	930	257	7.7	0.8	14	$(\pm45/0_2/\pm45/0_2/\pm45)_s$
8	908	245	6.6	0.8	12	$(\pm45/0_4/\pm45)_s$
9	882	234	5.5	0.8	10	$(\pm45/0_4)_s$
10	861	222	5.5	0.8	10	$(\pm45/0_4)_s$
11	875	211	5.5	0.8	10	$(\pm45/0_4)_s$
12	936	199	5.5	0.8	10	$(\pm45/0_4)_s$
13	1035	185	5.5	0.8	10	$(\pm45/0_4)_s$
14	1128	170	5.5	0.78	10	$(\pm45/0_4)_s$
15	1094	156	5.5	0.65	10	$(\pm45/0_4)_s$
16	1031	142	4	0.51	8	$(\pm45/0_3)_s$
17	956	130	4	0.33	8	$(\pm45/0_3)_s$
18	709	91	3	0.33	6	$(\pm45/0_2)_s$
19	543	61	3	0.33	6	$(\pm45/0_2)_s$
20	1037	44	2	0.2	4	$(\pm45/0)_s$

Table 8.
Maximum frequency design with discrete spanwise material grading.

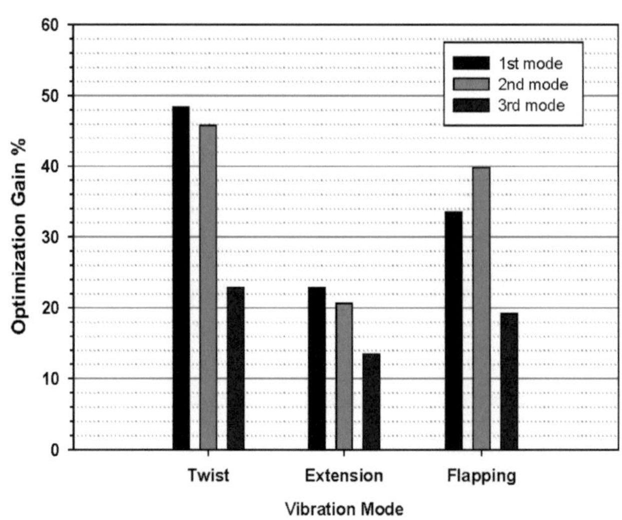

Figure 10.
Maximum frequency optimization gains.

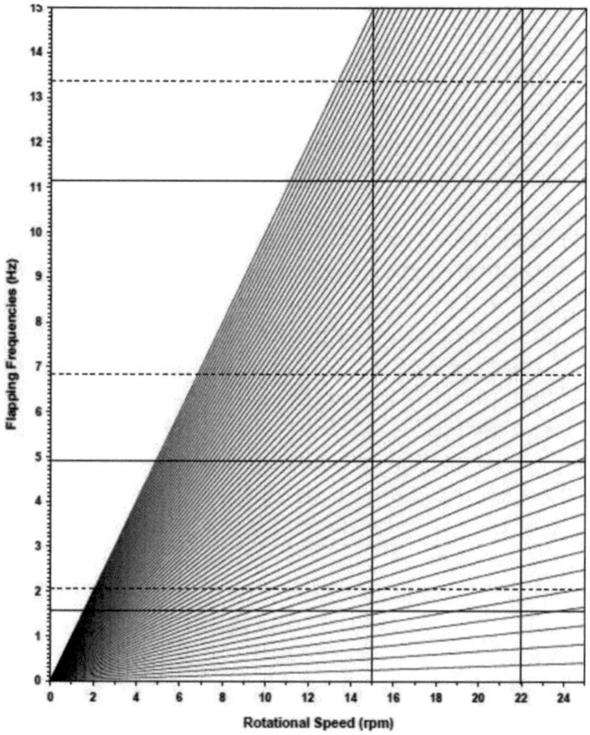

Figure 11.
Campbell diagram for baseline and maximum frequency designs. (———Baseline, --- max. $\sum_n$).

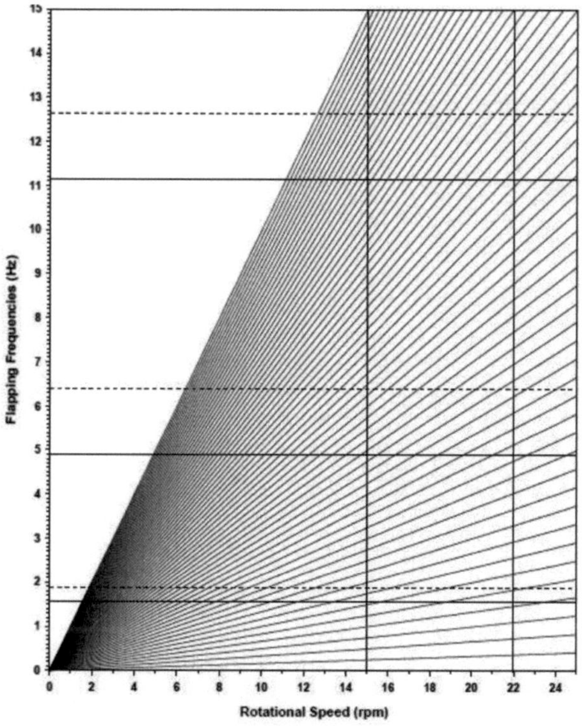

Figure 12.
Campbell diagram for baseline and frequency-placement designs. (——— baseline, ---- target).

constraint. **Table 8** presents the attained optimal solutions for the blade spar with discrete grading model.

Figure 10 shows the optimization gains for the different natural frequencies of the blade spar with maximum frequency designs. It is noticed that gains of 48%, 23%, and 34% have been achieved in fundamental natural frequencies for twist, extension, and flapping modes of vibration, respectively.

It can be concluded that a good blade design with maximum natural frequencies should have smaller wall thickness in the outboard portion of the blade spar. The maximum frequency optimization process recommends the segments located inboard to have higher wall thickness and higher fiber content. The spanwise variation in the height of each segment of the blade spar is always restricted by the airfoil envelope, which is the major obstacle in the optimization process.

When the wind turbine is operating, the rotating blades are the main source of vibration. The forcing frequencies are integer multiples of the rotational speed. A common way to represent natural frequencies and search for possible resonances is to plot the Campbell diagram as shown in **Figure 11**. The intersection of one of the radial lines with one of the system natural frequencies indicates a potential for resonance at the rotor speed corresponding to that point of intersection. The wind turbine considered in this investigation has two rotational speeds; they are 15 and 22 rpm. It is shown that for the baseline and maximum frequency designs, no resonance can occur at these operating speeds.

Segment no.	Segment length (mm)	Segment cross section (mm)		Fiber volume fraction	No. of layers	Layup
		Height	Thickness			
1	753	330	14.5	0.80	26	$(\pm45/0_5/\pm45/0_5/\pm45)_s$
2	948	318	12.9	0.80	24	$(\pm45/0_4/\pm45/0_5/\pm45)_s$
3	754	308	12.1	0.80	22	$(\pm45/0_4/\pm45/0_4/\pm45)_s$
4	787	298	12.3	0.80	22	$(\pm45/0_4/\pm45/0_4/\pm45)_s$
5	1008	284	11.9	0.80	22	$(\pm45/0_4/\pm45/0_4/\pm45)_s$
6	540	275	11.1	0.79	20	$(\pm45/0_3/\pm45/0_4/\pm45)_s$
7	460	271	10.8	0.79	20	$(\pm45/0_3/\pm45/0_4/\pm45)_s$
8	256	257	9.9	0.78	18	$(\pm45/0_7/\pm45)_s$
9	216	251	9.9	0.78	18	$(\pm45/0_8)_s$
10	186	241	10.9	0.71	20	$(\pm45/0_9)_s$
11	1431	244	7.8	0.74	14	$(\pm45/0_6)_s$
12	2563	210	6.4	0.80	12	$(\pm45/0_5)_s$
13	1930	185	6.5	0.80	12	$(\pm45/0_5)_s$
14	1536	165	6.4	0.80	12	$(\pm45/0_5)_s$
15	1374	147	5.6	0.80	10	$(\pm45/0_4)_s$
16	1310	125	6.4	0.43	12	$(\pm45/0_5)_s$
17	585	77	5.3	0.77	10	$(\pm45/0_4)_s$
18	185	51	10.9	0.72	20	$(\pm45/0_9)_s$
19	431	71	5.5	0.36	10	$(\pm45/0_4)_s$
20	1077	44	5.5	0.63	10	$(\pm45/0_4)_s$

Table 9.
Frequency-placement design with discrete spanwise material grading.

6.3.3 Frequency-placement design

In the minimal mass design, the natural frequencies of the optimized blade spar are constrained to be equal to those of the baseline design, while in maximum frequency optimization, the natural frequencies are entirely maximized with constraints imposed on its structural mass. If it is desired to have a blade spar with specific target frequencies, the frequency-placement optimization model has to be applied. The desired target frequencies used in the present analysis are selected to be 1.9, 6.4, and 12.65 Hz for the first three modes of flapping vibration, respectively. These frequencies are placed away from resonant frequencies at the operating speeds according to the Campbell diagram shown in **Figure 12**. The attained optimal solutions of the blade spar are given in **Table 9**, with the flapping frequencies placed at their prescribed target frequencies.

7. Finite element analysis

The finite element method (FEM) has been demonstrated as a powerful approach which can handle dynamic analysis of laminated composite structures. Nowadays, current commercial finite element software is capable of simulating nonlinearity of many engineering problems. Commercial software also come with advanced preprocessing and post-processing abilities [32] The preprocessing is just a way used for the data input, since the finite element method requires a large amount of data, while the post-processing is another way for presenting the results in the form of deformed shapes and contour maps. The core of the analysis is what occurs in between the two processes. In selecting a finite element analysis software, it is essential to regard the pre- and post-processor, which can directly affect the analysis speed and accuracy. One of the most appropriate finite element analysis software is *Femap* (finite element modeling and post-processing). *Femap* is an engineering analysis program developed by *Siemens PLM Software* that is used to build finite element models of complex engineering problems and view solution results [33]. *Femap* uses NX Nastran solver to analyze and solve finite element problems.

Femap is an advanced engineering simulation application for creating, editing, and importing/reusing mesh-centric finite element analysis models of complex products or systems. Femap provides powerful data-driven and graphical result visualization and evaluation.

The attained optimum design of the blade spar is modeled using FEM as a tapered beam with rectangular cross section. The root and tip cross sections are drawn using *Femap*, and then each line in the root is ruled to the corresponding line in the tip in order to generate the required surfaces as shown in **Figure 13**. The geometric model is divided into 20 segments with lengths according to the maximum frequency design given in **Table 8**.

Analysis	Natural frequencies (Hz)		
	1VBV	2VBV	3VBV
Analytical	2.09	6.87	13.30
FEM	2.17	7.04	14.49
Diff. %	4.7	2.4	8.2

Table 10.
Comparison between analytical and FEM modeling for the maximum frequency design.

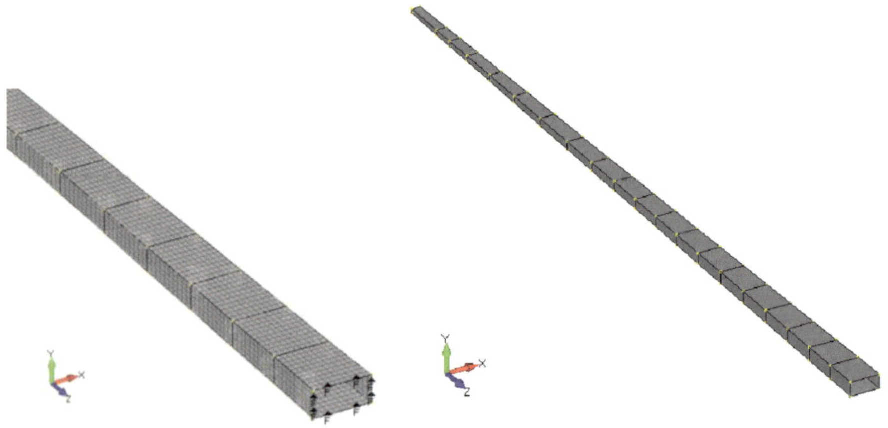

Figure 13.
Finite element model of a 20-segment cantilevered spar meshed with 10,000 elements.

Material properties, element type, and thickness of each segment are to be defined. According to **Table 8**, a number of six different orthotropic materials with different fiber volume fractions and properties of E_{11}, E_{22}, G_{12}, ν_{12}, and ρ are to be defined and input to *Femap*. A total of 20 different layups are to be defined due to the change in fiber orientation angles and thickness of each segment within the spar design. Furthermore, 20 element types are to be defined according to the corresponding material and layup properties.

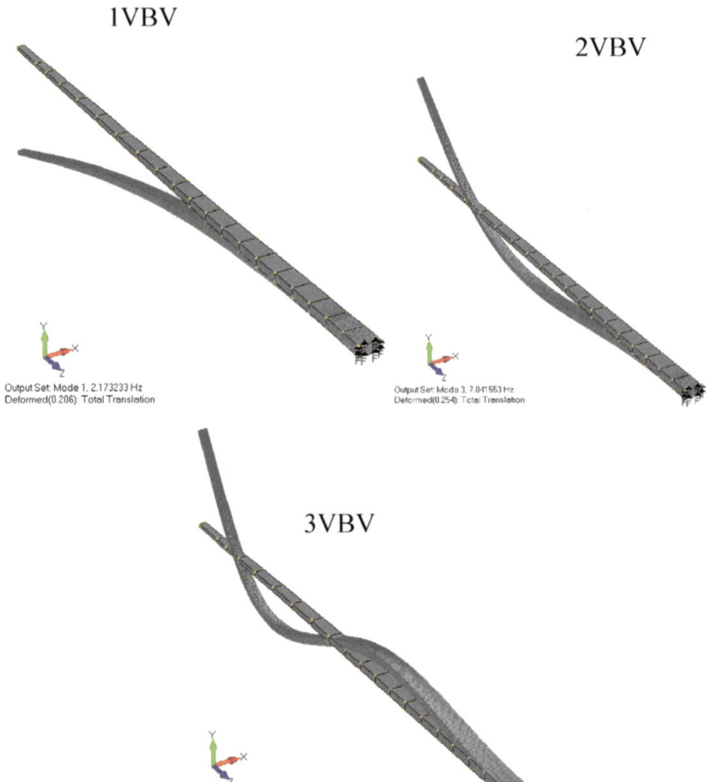

Figure 14.
Mode shapes and natural frequencies for vertical bending vibration of the blade spar with maximum frequency design.

The normal mode eigenvalues analysis using Lanczos modal analysis is used by *Femap* in order to define the different mode shapes and the associated natural frequencies of the optimized blade spar design. **Figure 14** shows the mode shapes with the associated natural frequencies for bending vibration of the optimized blade spar. **Table 10** shows a comparison between analytical and *Femap* natural frequencies of the optimized blade spar design.

8. Conclusions and recommendations

This chapter presents an optimization model for enhancing the dynamic performance of the spar beam of a wind turbine blade. Design variables include the cross-sectional dimensions and material properties variation along the spanwise direction. Three optimization strategies are developed and tested, including the minimal mass design, maximum frequency design, and frequency-placement design, by placing the frequencies at their target values to avoid large amplitudes and resonance occurrence. Side constraints are imposed on the design variables in order to avoid abnormal-shaped optimized configurations. Based on the fact that an exact dynamic analysis of uniform thin-walled beam segment is available and well established, the dynamic analysis of tapered blade spar has been obtained by applying the transfer matrix method to calculate the natural mode of vibrations. The proposed model deals with dimensionless quantities in order to be applicable to thin-walled beams with arbitrary dimensions. Results indicated that the optimization process leads to significant increase of natural frequencies of the optimized spar when compared to the reference or baseline design without mass penalty. Finite element model showed a good agreement with the analytical model developed in this study with a variation of up to 10%. The main conclusions that can be revealed from the present work are:

1. Tapered multiple-segment spar with spanwise material grading gives natural frequencies higher than that of the reference design. However, it is proved that maximization of the fundamental frequency alone does not guarantee maximization of the other higher frequencies. Higher frequencies have been found to have many local minima and maxima in the defined design space.

2. There are optimum design variables of each segment such as the length, height, wall thickness, and fiber content at which the structural dynamic performance can be enhanced. Good designs favor minimum wall thickness and higher fiber volume fraction.

Finally, the analytical model formulated in this chapter can be extended and applied to study the forced dynamic response of a wind turbine blade. Other cross-sectional types of the blade spar, such as D-shape spar, can be considered, and the effects of blade twist, shear deformation, and rotary inertia are to be considered in future studies.

Appendix

A. Failure criteria of fibrous composite materials

The fracture processes induced in fibrous composite materials depend upon the nature of constituents, the architecture of the laminate, and the type of mechanical

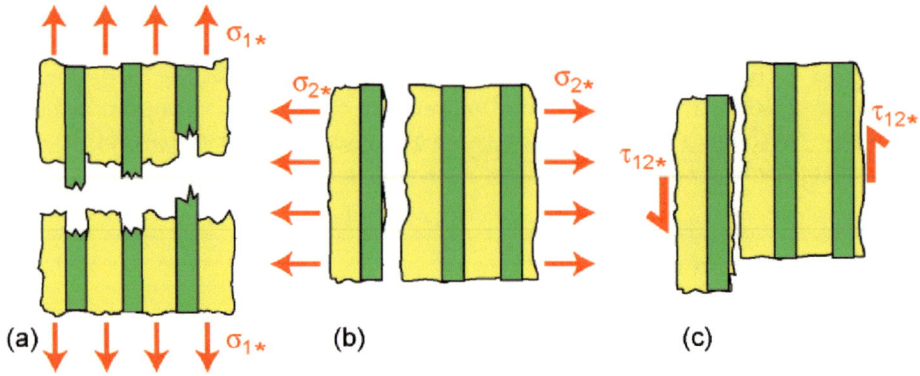

Figure A.1.
Schematic of the fracture of a unidirectional fibrous composite at critical values of (a) axial, (b) transverse, and (c) shear stresses.

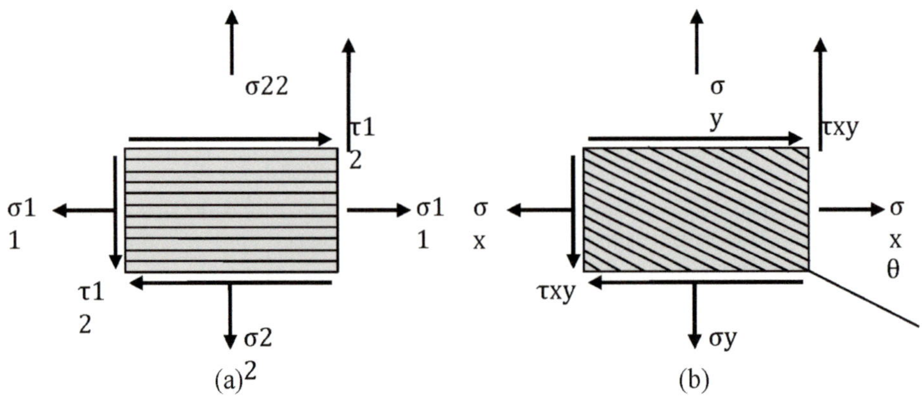

Figure A.2.
Lamina under loading condition of (a) principal axes and (b) general axes.

loading imposed to the laminate. The rupture of fibrous composite materials is the result of one or combined effect of fiber fracture, transverse fracture in matrix, longitudinal fracture in matrix, fracture of fiber-matrix interface, and delaminations [34] (refer to **Figure A.1**).

Failure theories for composites have been proposed by extending and adapting isotropic failure theories to account for the anisotropy in stiffness and strength of the composites. One of the first fracture criteria applied to an anisotropic materials was introduced by Tsai-Hill theory [26], which is applied to unidirectional lamina under principal axis in-plane loading condition as shown in **Figure A.2 (a)**. The theory states that no fracture will occur in the lamina if $\alpha_T \leq 1$ such that:

$$\alpha_T = \left(\frac{\sigma_{11}}{\sigma_{11r}}\right)^2 + \left(\frac{\sigma_{22}}{\sigma_{22r}}\right)^2 - \frac{\sigma_{11}\sigma_{22}}{\sigma_{11r}^2} + \left(\frac{\tau_{12}}{\tau_{12r}}\right)^2 \quad \text{(A.1)}$$

where the subscript r refers to rupture strength of the material.

Rupture strengths for fibrous composite materials are given by the following approximate relations [34]:

$$\begin{aligned} \sigma_{11r} &= \sigma_{11fr}\left(V_f + V_m E_m / E_{11f}\right) \\ \sigma_{22r} &= \sigma_{mr}\left(1 - \sqrt{4V_f/\pi}\right) \\ \tau_{12r} &= \tau_{mr} \end{aligned} \quad \text{(A.2)}$$

Subscripts *f* and *m* refer to fiber and matrix materials, respectively.

For a lamina under general loading condition as shown in **Figure A.2 (b)**, the principal stresses are given by the following equation [4, 34]:

$$\begin{Bmatrix} \sigma_{11} \\ \sigma_{22} \\ \tau_{12} \end{Bmatrix} = \begin{bmatrix} c^2 & s^2 & -2cs \\ s^2 & c^2 & 2cs \\ cs & -cs & (c^2 - s^2) \end{bmatrix} \begin{Bmatrix} \sigma_x \\ \sigma_y \\ \tau_{xy} \end{Bmatrix} \tag{A.3}$$
$$c = cos\theta s = sin\theta$$

The Tsai-Hill failure theory is expressed in terms of a single criterion instead of the multiple subcriteria required in the maximum stress and maximum strain theories. The theory allows for considerable interaction among the stress components. One disadvantage, however, is that it does not distinguish directly between tensile and compressive strengths. The strength must be specified and used according to the given state of stress.

B. Sequential quadratic programming

Sequential quadratic programming (SQP) is one of the mainly developed and perhaps one of the best techniques of optimization [29]. The method has a theoretical base that is related to the solution of a group of nonlinear equations using Newton's method and the derivation of concurrent nonlinear equations using Karush-Kuhn-Tucker (*KKT*) conditions to the Lagrangian of the constrained optimization problem. The fundamental idea of SQP is to model the optimization problem at the present iterate x_k by a quadratic programming subproblem and to employ the minimizer of this subproblem to identify a new iterate x_{k+1} [30]. The challenge is to design the quadratic subproblem so that it yields a good step for the constrained optimization problem and so that the overall SQP algorithm has good convergence properties and good practical performance. Possibly the simplest derivation of SQP methods views them as an application of Newton's method to the *KKT* optimality conditions for the optimization problem. If the problem is a so-called convex programming problem, that is, $f(\boldsymbol{X})$ and $G_i(\boldsymbol{X})$, $i = 1, \dots, m$, are convex functions, then the *KKT* equations are both necessary and sufficient for a global solution point. The Kuhn-Tucker equations can be stated as:

$$\begin{aligned} &\nabla f(x) + \sum_{i=1}^{m} \lambda_i . \nabla G_i(x) = 0 \\ &\lambda_i . G_i(x) = 0, i = 1, \dots, m_e \\ &\lambda_i \geq 0, i = m_e + 1, \dots, m \end{aligned} \tag{B.1}$$

The first equation describes a canceling of the gradients between the objective function and the active constraints at the solution point. For the gradients to be canceled, Lagrange multipliers (λ_i, $i = 1, \dots, m$) are necessary to balance the deviations in magnitude of the objective function and constraint gradients. Because only active constraints are included in this canceling operation, constraints that are not active must not be included in this operation and so are given Lagrange multipliers equal to 0. This is stated implicitly in the last two Kuhn-Tucker equations.

B.1 Optimization using MATLAB

MATLAB is popular software that is used for the solution of a variety of scientific and engineering problems. The specific toolbox of interest for solving

Argument	Description
fun	The function to be minimized
x0, lb., and ub	Starting, lower, and upper boundaries of design variables
nonlcon	The function that computes the nonlinear inequality constraints "c(x) < 0" and the nonlinear equality constraints "ceq(x) = 0"
fval	Value of the function "fun" at x
exitflag	Integer identifying the reasons causing the algorithm terminated
grad	Gradient at x
hessian	Hessian at x
lambda	Structure containing the Lagrangian multipliers at the solution x
output	Structure containing information about the optimization process such as number of function evaluations and iterations taken and the used optimization technique

Table B.1.
Arguments related to fmincon *function [35].*

optimization problems is called the optimization toolbox [35]. MATLAB optimization toolbox contains a library of programs or m-files, which can be used for the solution of different optimization problems. A commonly implemented function denoted by *fmincon* is applied to most constrained objective functions. It is the most suitable function to the optimization problem of this investigation, since it uses sequential quadratic programming (SQP) method.

fmincon starts at "x0" and attempts to find a minimizer "x" of the objective function described in the m-file named "fun" subject to the linear inequalities "$A \times x \leq b$" and the linear equalities "$Aeq \times x = beq$." If there are no linear equalities or inequalities, A, b, Aeq, and beq are replaced with "[]." *fmincon* can define lower and upper bounds on the design variables in "x" so that the solution is always in the range "lb ≤ x ≤ ub." *fmincon* can subject the minimization process to the nonlinear inequalities "c(x)" or equalities "ceq(x)" defined in the m-file constraint function named "nonlcon." *fmincon* optimizes such that "c(x) ≤ 0" and "ceq(x) = 0." *fmincon* minimizes with the optimization options involved in the structure "options." **Table B.1** defines all input and output arguments related to *fmincon* optimization function.

C. Applied loads and stress analysis

The distributed load vectors are expressed in the undeformed coordinates (see **Figure C.1**), as follows:

$$\text{Distributed forces}: \underline{P} = \underline{P}_A + \underline{P}_I + \underline{P}_G + \underline{P}_D \tag{C.1}$$

$$\text{Distributed moments}: \underline{q} = \underline{q}_A + \underline{q}_I + \underline{q}_G + \underline{q}_D \tag{C.2}$$

where the subscripts A, I, G, and D refer to the aerodynamic, inertial, gravitational, and damping contributions, respectively. The aerodynamic forces $\underline{P}_A$ and moments $\underline{q}_A$ can be obtained using the quasi-steady blade-element strip theory [1, 2].

Figure C.2 shows the velocity triangle and the coordinate system used for computing aerodynamic loads. V_{yr} and V_{zr} are the tangential and axial velocity components, respectively. The resultant velocity V_r can be calculated from:

$$V_r \simeq \sqrt{V_{yr}^2 + V_{zr}^2} = -V_{yr}\sqrt{1+\left(\frac{V_{zr}}{V_{yr}}\right)^2} \tag{C.3}$$

The distributed lift and drag force vectors on an arbitrary airfoil section are given by the aerodynamic formulas (see **Figure C-2**):

$$\begin{aligned} \text{Lift } \underline{L} &= \frac{1}{2}\,\rho_a V_r^2\, C\, C_L\left(\sin\,\varphi_i\hat{j}' + \cos\,\varphi_i\hat{k}'\right) \\ \text{Drag } \underline{D} &= \frac{1}{2}\,\rho_a V_r^2\, C\, C_D\left(-\cos\,\varphi_i\hat{j}' + \sin\,\varphi_i\hat{k}'\right) \end{aligned} \tag{C.4}$$

where ρ_a = air density; C = local chord at spanwise location r; $C_L = C_L(\alpha) = C_{L\alpha}\alpha$, lift coefficient; $C_D = C_D(\alpha)$, drag coefficient; $C_{L\alpha}$= lift – curve slope; α = angle of attack; φ_i = inflow angle = $\theta_B + \alpha$.

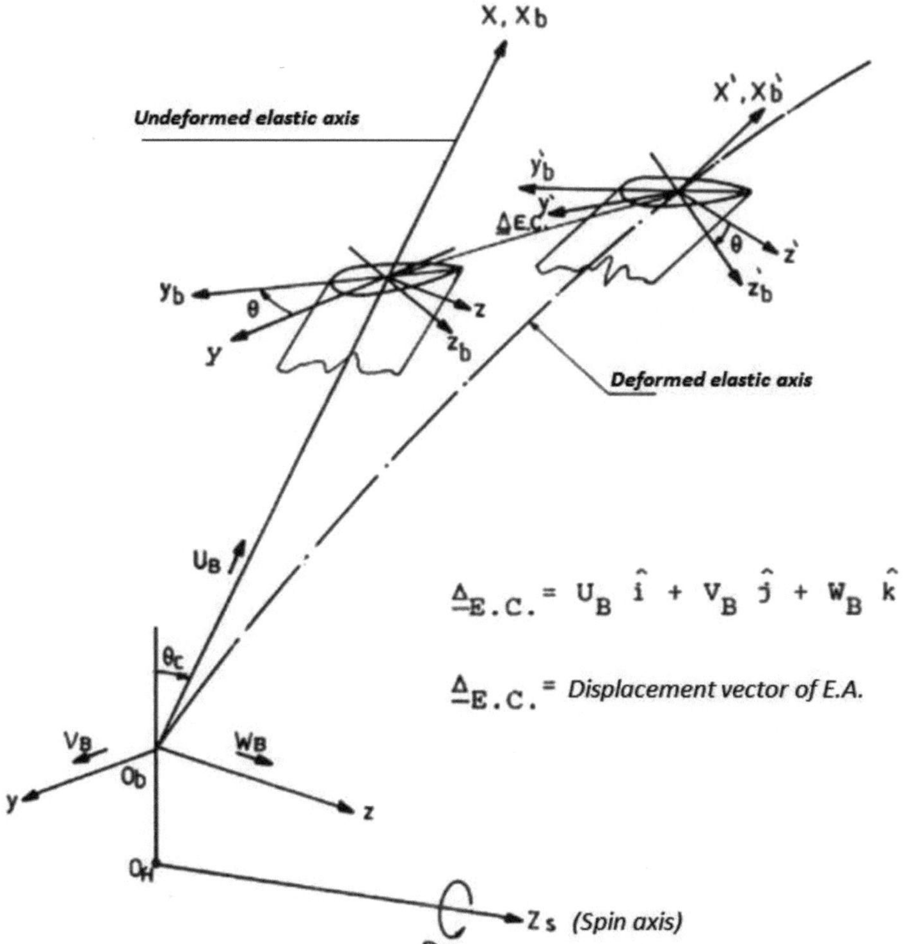

Figure C.1.
Deformed and undeformed configurations of a blade segment.

Expressed in the undeformed system, the aerodynamic force vector per unit length of the blade is therefore given as:

$$\underline{P}_A = \underline{L} + \underline{D} = \mathrm{P}_{\mathrm{xA}}\hat{i} + \mathrm{P}_{\mathrm{yA}}\hat{j} + \mathrm{P}_{\mathrm{zA}}\hat{k} \qquad \text{(C.5)}$$

The distributed inertia loads per unit blade span are obtained by applying D'Alembert's principle as follows:

$$\begin{aligned} &\text{Inertial force vector}: \underline{P}_{\mathrm{IB}} = -\iint \rho_B \ddot{\underline{r}}_B \,\mathrm{dy\,dz} \\ &\text{Inertial moment vector}: \underline{q}_{\mathrm{IB}} = -\iint \rho_B \left[\left(\underline{r}'_B - r'_{E.C} \right) \times \ddot{\underline{r}}_B \right] \mathrm{dy\,dz} \end{aligned} \qquad \text{(C.6)}$$

where ρ_B is the blade material mass density, $\ddot{\underline{r}}_B$is the acceleration vector, and $\underline{r}'_B$ and $\underline{r}'_{E.C}$ are the position vectors of an arbitrary point and the elastic center of the blade cross section, respectively.

Next, considering the equilibrium of a differential element of the deformed blade, the equilibrium equations expressed in the rotating (xyz) axes are:

$$\begin{aligned} &\text{Force equilibrium}: \quad \frac{\partial \underline{F}}{\partial x} + \underline{P} = \underline{0} \\ &\text{Moment equilibrium}: \quad \frac{\partial \underline{M}}{\partial x} + \hat{i}' \times \underline{F} + \underline{q} = \underline{0} \end{aligned} \qquad \text{(C.7)}$$

$$\begin{aligned} &\text{Internal force vector}: \quad \underline{F} = \mathrm{F}_x\hat{i} + \mathrm{F}_y\hat{j} + \mathrm{F}_z\hat{k} \\ &\text{Internal moment vector}: \quad \underline{M} = \mathrm{M}_x\hat{i} + \mathrm{M}_y\hat{j} + \mathrm{M}_z\hat{k} \end{aligned} \qquad \text{(C.8)}$$

A wind turbine blade cross section is composed of thin-walled, closed, single or multicellular composite beams. The periphery of each beam cross section is assumed to be constructed of flat composite laminates as shown in **Figure C.5**. The stress-strain relations of the composite laminates which discretize each cross section are computed using the classical laminate theory *CLT*. Although each laminate is

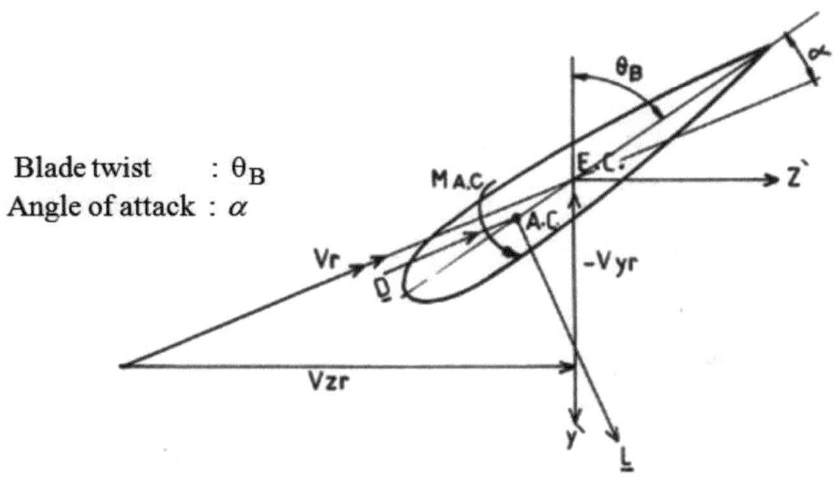

Figure C.2.
Velocity triangle of an arbitrary airfoil section in the deformed state.

actually an assembly of multiple layers with different constitutive properties, *CLT* is used to calculate a set of effective stiffness coefficients that allows a composite laminate to be treated as a single structural element [12]. Therefore the blade cross section may be considered to be composed of discrete sections of homogenous materials.

Generally, for a heterogeneous composite section shown in **Figure C.3** and **C.4**, the modulus-weighted cross-sectional properties are defined as [12]:

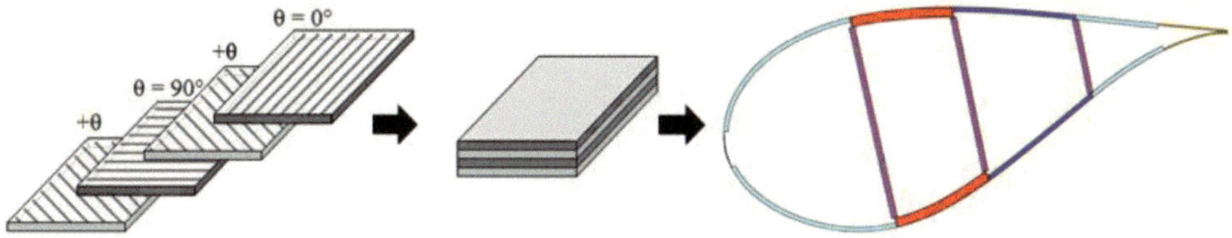

Figure C.3.
Blade cross section with discretized composite laminated plates.

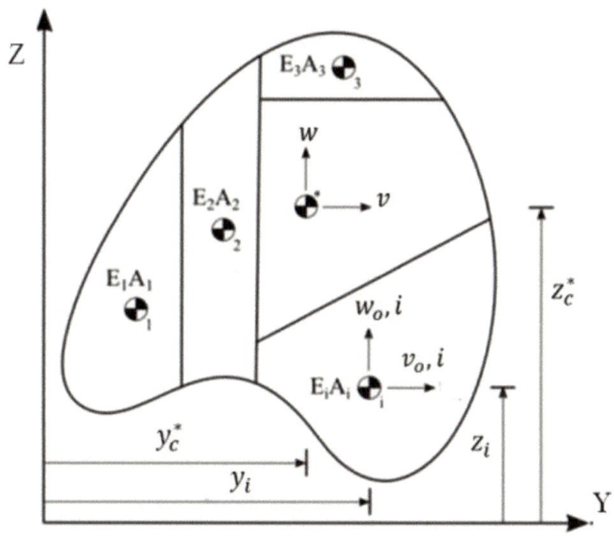

Figure C.4.
Cross section for a heterogeneous composite beam.

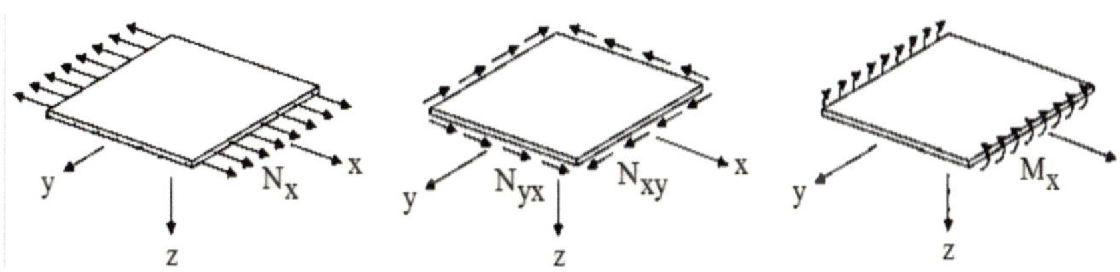

Figure C.5.
Resultant loads and moments applying to a composite laminate.

$$
\begin{aligned}
A_c &= \frac{1}{E_{ref}} \sum_{i=1}^{n} E_i A_i \\
y_c &= \frac{1}{E_{ref} A_c} \sum_{i=1}^{n} E_i A_i y_i \\
z_c &= \frac{1}{E_{ref} A_c} \sum_{i=1}^{n} E_i A_i z_i \\
I_y &= \frac{1}{E_{ref}} \sum_{i=1}^{n} E_i \left(I_{v_o,i} + A_i z_i^2 \right) \\
I_z &= \frac{1}{E_{ref}} \sum_{i=1}^{n} E_i \left(I_{w_o,i} + A_i y_i^2 \right) \\
I_{yz} &= \frac{1}{E_{ref}} \sum_{i=1}^{n} E_i \left(I_{vw_o,i} + A_i y_i z_i \right)
\end{aligned}
\tag{C.9}
$$

where E_{ref} is a reference modulus of elasticity, (y_i, z_i) denotes the geometric centroid of each discrete element of the cross section, and $(v_{o,i}, w_{o,i})$ denotes the principal axes of each discrete element.

The parallel axes theorem can be applied to compute the second moments of area about the cross-sectional principal axes as follows:

$$
\begin{aligned}
I_v &= I_y - A_c (z_c)^2 \\
I_w &= I_z - A_c (y_c)^2 \\
I_{vw} &= I_{yz} - A_c y_c z_c
\end{aligned}
\tag{C.10}
$$

Once the global cross-sectional properties are computed using the method of modulus-weighted properties, the effective axial stress applied to the blade cross section can be given by:

$$
\sigma_x(y, z) = \frac{F_x}{A_c} - \frac{M_z I_v + M_y I_{vw}}{I_v I_w - I_{vw}^2} (y - y_c) + \frac{M_y I_w + M_z I_{vw}}{I_v I_w - I_{vw}^2} (z - z_c) \tag{C.11}
$$

Finally, by converting the distribution of the effective beam stresses into equivalent in-plane distributed loads on the flat laminates composing the cross section, as shown in **Figure C.5**, the lamina-level strains and stresses can be computed using the *CLT*.

Author details

Gerges Edwar Mehanny Beshay[1]* and Karam Yousef Maalawi[2]

1 Faculty of Engineering (Shoubra), Department of Mechanical Engineering, Benha University, Cairo, Egypt

2 Department of Mechanical Engineering, National Research Centre, Dokki, Giza, Egypt

*Address all correspondence to: gerges.beshay@feng.bu.edu.eg

References

[1] Al-Bahadly I. Wind Turbines. Croatia: IntechOpen; 2011. Available from: http://www.intechopen.com/books/wind-turbines. ISBN: 978-953-307-221-0

[2] Gasch R, Twele J. Wind Power Plants. New York: Springer; 2012. ISBN: 978-3-642-22938-1

[3] Gay D, Hoa SV. Composite Materials: Design and Applications. New York: CRC Press; 2007

[4] Daniel I, Ishai O. Engineering Mechanics of Composite Materials. 2nd ed. New York: Oxford University Press; 2006. pp. 432. ISBN: 978-0195150971

[5] Suresh S, Mortensen A. Fundamentals of Functionally Graded Materials. Cambridge University Press; 1998

[6] Birman V, Byrd WL. Modeling and analysis of functionally graded materials and structures. Applied Mechanics Reviews. 2007;**60**(5):195-216

[7] Maalawi KY, Negm HM. Optimal frequency design of wind turbine blades. Journal of Wind Engineering and Industrial Aerodynamics. 2002; **90**(8):961-986

[8] Maalawi KY. A model for yawing dynamic optimization of a wind turbine structure. International Journal of Mechanical Sciences. 2007;**49**:1130-1138

[9] Librescu L, Maalawi K. Material grading for improved aeroelastic stability in composite wings. Journal of Mechanics of Materials and Structures. 2007;**2**(7):1381-1394

[10] Chen K-N, Chen P-Y. Structural optimization of 3 MW wind turbine blades using a two- step procedure. International Journal for Simulation and Multidisciplinary Design Optimization. 2010;**4**:159-165. DOI: 10.1051/ijsmdo/2010020

[11] Maalawi KY, Badr MA. Frequency optimization of a wind turbine blade in pitching motion. Journal of Power and Energy, Proceedings of the Institution of Mechanical Engineers, Part A. 2010; **224**:545-554. DOI: 10.1243/09576509JPE907

[12] Sale D, Aliseda A, Motley M, Li Y. Structural optimization of composite blades for wind and hydrokinetic turbines. In: Proceeding of 1st Marine Energy Technology Symposium, METS 2013. Washington, DC, April 10–11. 2013

[13] Chehouri A, Younes R, Ilinca A, Perron J, Lakiss H. Optimal design for a composite wind turbine blade with fatigue and failure constraints. Transactions of the Canadian Society for Mechanical Engineering, CSME-16. 2015;**39**(2):171-186

[14] Maalawi KY. Dynamic optimization of functionally graded thin-walled box beams. International Journal of Structural Stability and Dynamics. 2017; **17**(9):1-24

[15] Armanios EA, Badir AM. Free vibration analysis of anisotropic thin-walled closed-sections beams. American Institute of Aeronautics and Astronautics (AIAA). 1995;**33**:1905-1910

[16] Dancila DS, Armanios EA. The influence of coupling on the free vibration of anisotropic thin-walled closed-section beams. International Journal of Solids and Structures. 1998; **35**:3105-3119

[17] Durmaz S, Kaya MO. Free vibration of an anisotropic thin-walled box beam under bending-torsion coupling.

In: Proceedings of 3rd International Conference On Integrity, Reliability and Failure. 2009

[18] Shadmehri F, Haddadpour H, Kouchakzageh M. Flexural-torsional behaviour of thin-walled composite beams with closed cross-section. Thin Walled Structures. 2007;**45**:699-705

[19] Phuong T, Lee J. Flexural-torsional behavior of thin-walled composite box beams using shear-deformable beam theory. Engineering Structures. 2008; **30**:1958-1968

[20] Piovan MT, Filipicha CP, Cortínez VH. Coupled free vibration of tapered box beams made of composite materials. Mecánica Computacional. 2006;**25**:1767-1779

[21] Kargarnovin MH, Hashemi M. Free vibration analysis of multilayered composite cylinder consisting fibers with variable volume fraction. Composite Structures. 2012;**94**:931-944

[22] Liu Y, Shu DW. Free vibration analysis of exponential functionally graded beams with a single delamination. Composites Part B Engineering. 2014;**59**:166-172

[23] Whitney JM. Elastic moduli of unidirectional composites with anisotropic filaments. Journal of Composite Materials. 1967;**1**:188-194

[24] Meirovitch L. Principles and Techniques of Vibrations. Englewood Cliffs, NJ: Prentice-Hall; 1997

[25] Mihail B, Valentin C, Costin D. A transfer matrix method for free vibration analysis of Euler-Bernoulli beams with variable cross section. Journal of Vibration and Control. 2014: 1-12. DOI: 10.1177/1077546314550699

[26] Halpin JC, Tsai SW. Effect of environmental factors on composite materials. In: AFML-TR. Vol. 67. 1969. pp. 1-53

[27] Rao S, Singiresu. Engineering Optimization: Theory and Practice. John Wiley & Sons Inc; 2009

[28] Maalawi K, Badr M. Design optimization of mechanical elements and structures: A review with application. Journal of Applied Sciences Research. 2009;**5**(2):221-231

[29] Fletcher R. The sequential quadratic programming method. In: Nonlinear Optimization. Springer; 2010. pp. 165-214

[30] Boggs PT, Tolle JW. Sequential quadratic programming for large-scale nonlinear optimization. Journal of Computational and Applied Mathematics. 2000;**124**(1):123-137

[31] Bir GS, Oyague F. Estimation of blade and tower properties for the gearbox research collaborative wind turbine. In: Technical Report NREL/TP-500-42250. U.S. Department of Energy: National Renewable Energy Laboratory; November 2007

[32] Khennane A. Introduction to Finite Element Analysis Using MATLAB and Abaqus. Boca Raton: CRC Press; 2013

[33] Available from: http://www.plm.automation.siemens.com/en_us/products/velocity/femap/features/

[34] Bryan H. Engineering Composite Materials. The Institute of Materials; 1999

[35] Grace A, Branch MA, Coleman T. Optimization Toolbox for use with MATLAB. The Math Works Inc.; 1999

Chapter 5

Aerodynamic, Structural and Aeroelastic Design of Wind Turbine Blades

Mohamed Abdou Mahran Kasem

Abstract

With the negative impact of conventional energy resources that have been used worldwide, there is a demand for using other resources such as wind energy. Tons of researches have been applied around the globe on the process of designing and manufacturing wind energy conversion systems. In the present chapter, we are concentrating on wind turbine blades' structural design process. The structural design of a wind turbine blade includes defining the wind turbine loads, selecting a suitable material, creating a structural model, and solving the model using the finite element method. This process will be repeated several times until a final design is achieved. The present chapter includes a discussion on the finite element method and wind turbine aeroelasticity.

Keywords: wind turbine blades, composite structures, finite element method, aeroelasticity

1. Introduction

With the increase in clean energy demand, wind turbine power and subsequently its size increased dramatically. The power generated from wind turbine increased from approximately 30 GW in 2001 to about 600 GW around the globe as depicted in **Figure 1**.

While the total power capacity in Europe increased from about 70 GW to 180 GW in the period from 2008 to 2018, as shown in **Figure 2**, in the United States, the wind power generated from wind turbines increased from 0 to 100 GW in the period from 2000 to 2019 [1].

This increase in wind turbine size makes it important to efficiently design wind turbine structure. A blade structure must be stiff enough, so it does not fail due to wind turbine loads, and at the same moment it should be designed in lightweight and low cost. In the present chapter, we discuss the general structural design process for a wind turbine blade for maximum strength to weight ratio. This process starts by calculating the wind turbine loads, selecting a suitable material, creating a structural model, and applying the finite element method for predicting aeroelastic stability boundaries.

IntechOpen

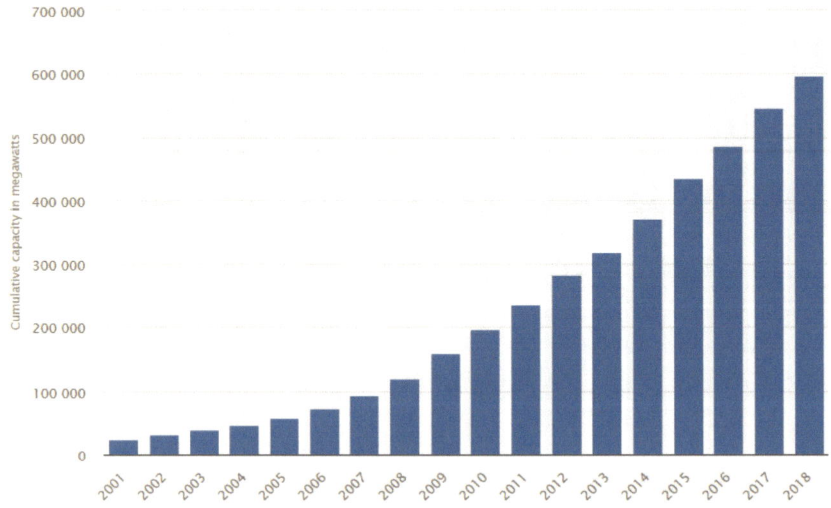

Figure 1.
Global wind power capacity from 2001 to 2018 [1].

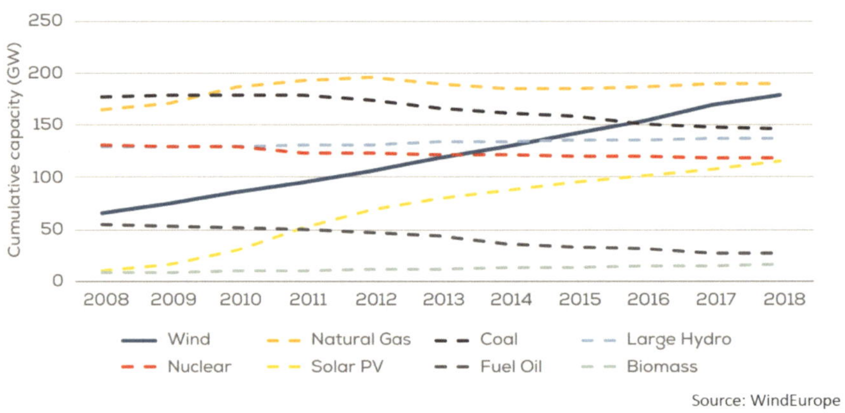

Figure 2.
The European Union's total power generation [2].

2. Blade aerodynamic loads

A wind turbine structure must be stiff enough to stand for the flow around. For this reason, the first step in the structure design process is to determine the aerodynamic loads applied to a wind turbine blade. There are mainly two methods for predicting a wind turbine aerodynamic load: blade element momentum (BEM) and computational fluid dynamics (CFD). The BEM can be classified as an analytical method that is fast and accurate. On the other hand, the CFD analysis is a numerical method which is based on numerical and empirical approximations. In general, the BEM is preferred for initial design and load estimation, while the CFD simulation is preferred for detailed design. Designers usually start using BEM in preliminary design and end up using the CFD simulation for detailed analysis and design. A comparison between the BEM and CFD analyses is listed in **Table 1**.

Despite the simplicity of the BEM, it can obtain accurate results as the CFD analysis [3]. However, the BEM is not preferred for detailed simulation. There are software available for both the BEM and CFD analyses, such as wind turbine design

BEM	CFD
Analytical method	Numerical method
Simple to be derived and used	Relatively complex in its formulation and application
Has relatively low computational time	Has relatively high computational time
In general, it does not consider 3D effects (just some corrections such as tip losses)	Depicts 3D effects in detail
Does not consider turbulence effects	Considers turbulence effects
Recommended for preliminary design	Recommended for detailed simulation

Table 1.
Comparison between the BEM and CFD methods.

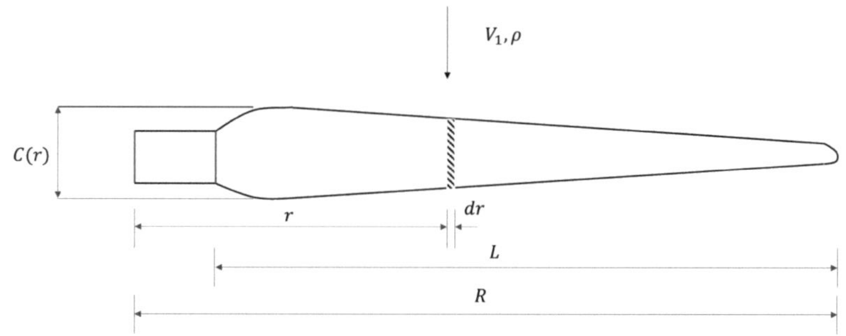

Figure 3.
BEM model.

and simulation (Q-blade) that is based on the BEM [4] and ANSYS Fluent that is based on CFD simulation [5]. For more details about ANSYS CFD simulation, we refer the reader to ANSYS Fluent Theory Guide [6]. In the following section, we will summarize the BEM method.

2.1 Blade element momentum theory

The BEM is considered as a simple and fast method in calculating the applied aerodynamic loads on wind turbine blade. The blade is divided into N radial segments; each segment experiences different chord (C), twist angle (β), and tangential speed (Ωr), as shown in **Figure 3**.

In **Figure 3**, V_1 denotes the upstream velocity, ρ is the flow density, R is the rotor radius, C represents the blade chord, and L denotes the blade length. r is the radial position of the rotor section, and dr defines the blade section width. In the present analyses, we know exactly the blade geometry (chord distribution, angle of twist, and airfoil characteristics), and our objective is to determine the aerodynamic loads applied to the blade (lift and drag).

In BEM, we calculate the forces on each blade element, and then the total forces over the blade can be calculated using summation. The lift force and blade angles are defined in **Figure 4**, which **Figure 4** shows the triangle of in-plane velocity (no subscript), triangle of upstream velocity (subscript 1), and triangle of downstream velocity (subscript 3) [7], where, ϕ is the relative velocity angle.

To simplify the process of calculating blade loads using BEM, we describe it as follows:

1. Start by

$$\phi = \phi_1 = \tan^{-1}\left(\frac{V_1}{\Omega r}\right) = \tan^{-1}\left(\frac{R}{\lambda r}\right) \tag{1}$$

where λ is the tip speed ratio.

2. Set the relative flow speed upper and lower limits:

$$\sin\phi_{max} = \frac{Z\sqrt{1-\frac{r2}{R}}}{2\pi\,\frac{r}{R}}, \text{and } \sin\phi_{min} = \sin\left(\frac{2}{3}\phi_1\right) \tag{2}$$

where Z represents the number of rotor blades.

3. Calculate the angle of attack from $\alpha_A = \phi - \beta$; then, one can obtain the section lift and drag coefficient from the airfoil characteristics.

4. Calculate the parameter $x = \sin\phi$.

If $x < \sin\phi_{min}$, use the Glauert correction
$\rightarrow x = 0.25\,\sin\phi_{min}\sqrt{9-2y^2+9y^4}, y = \frac{\sin\phi}{\sin\phi_{min}}$, or
if $x > \sin\phi_{max}$, use the Prandtl correction $\rightarrow x = \sin\phi_{max}$.

5. Solve the BEM nonlinear equation to obtain ϕ:

$$\phi = \phi_1 - \tan^{-1}\left(\frac{C_l}{\frac{8\pi r}{ZC}x + C_d}\right) \tag{3}$$

6. Repeat steps from 1 to 5 till convergence.

7. After convergence, calculate the relative velocity w:

$$w = w_1\cos(\phi_1 - \phi)\frac{\frac{8\pi r}{Z}x}{\frac{8\pi r}{Z}x + C\,C_d} \tag{4}$$

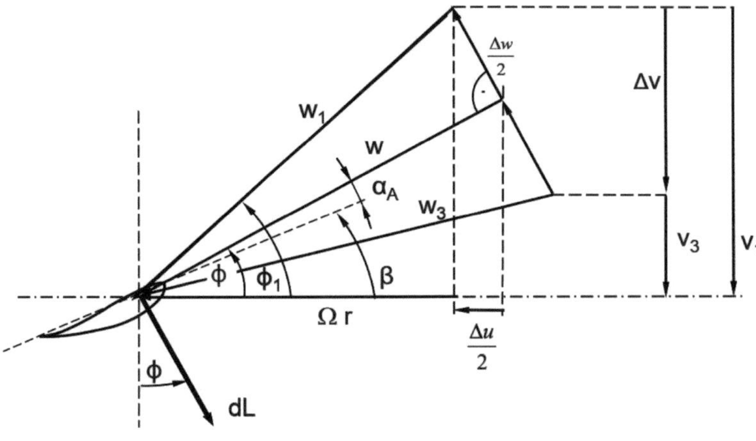

Figure 4.
Blade element force and angles.

8. Calculate the aerodynamic forces (lift and drag forces):

$$dL = \frac{\rho}{2} w^2 C\, dr\, C_l \tag{5}$$

$$dD = \frac{\rho}{2} w^2 C\, dr\, C_d \tag{6}$$

- The thrust force can be calculated from the equation:

$$F_n = Z \sum_{1}^{N} (dL \ \cos\phi + dD \ \sin\phi) \tag{7}$$

where N represents the number of blade elements.

- The blade tangential (circumferential force)

$$F_z = \sum_{1}^{N} (dL \ \sin\phi - dD \ \cos\phi) \tag{8}$$

- Blade torque

$$M = \sum_{1}^{N} (dL \ \sin\phi - dD \ \cos\phi)\, r \tag{9}$$

Blade power

$$P = \Omega M \tag{10}$$

A schematic for a blade aerodynamic pressure is shown in **Figure 5**. It is seen that the maximum force occurs at the blade root and the minimum at the tip. For this reason, wind turbine blades are usually designed with taper in which the airfoil thickness increases toward the blade root. This property makes the blade structure stiffer at the root and lighter at the tip. For this reason, it is recommended to also design the wind turbine structure with taper. In other words, ribs, spars, and skins have thickness at the blade root higher than the thickness at the tip.

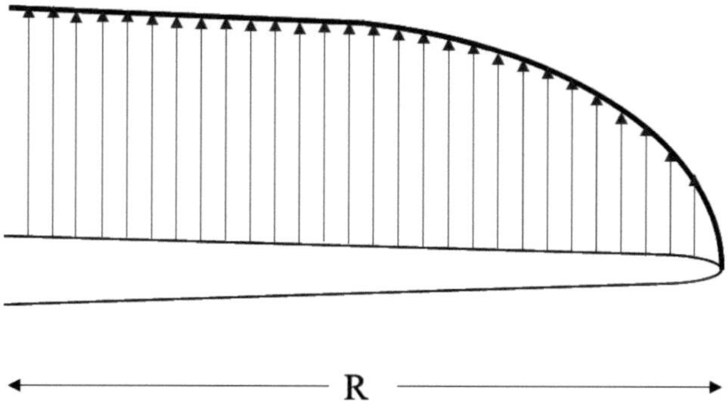

Figure 5.
Schematic for a blade aerodynamic load.

3. Material selection

Wind turbines have been made from different materials such as wood, aluminum, and composites. Modern wind turbines are usually made from composites such as carbon fibers and fiber glass. The wide use of composite materials turns to its relative high stiffness to weight ratio in addition to its ability to form complex shapes. Composites are found to be efficient with the large increase in wind turbines' size and capacities. They also can be tailored to satisfy different stiffness and weight requirements. Composite fiber can be used in different orientation to improve the blade directional stiffness in addition to bending and torsional rigidity. **Figure 6** shows the layup process for manufacturing a 6-m-diameter wind turbine that was designed and manufactured at Cairo University laboratories.

Two important things have to be considered when selecting composites: first, selecting the proper fiber direction for a blade structure, and second, insuring that the final product (blade structure) has the same material properties as it was desired in the design process. The former point can be overcome by applying a proper optimization process to select the best laminate configuration for maximizing the blade stiffness to weight ratio, while the latter issue can be resolved by testing the layup configuration after manufacturing some samples to make sure they have the same desired properties plus making experimental tests and measurements to compare between what was designed and what was manufactured.

There are an infinite number of orientations for composite structures. A composite layer can be oriented in any direction. However, it is important when selecting a blade laminate configuration to consider 0, 90, and ±45° angles for the blade skin. These directions are the most important fiber directions to increase the longitudinal, directional, torsional, and bending stiffnesses of a blade (**Figure** 7).

Figure 6.
Fiberglass/epoxy blade layup.

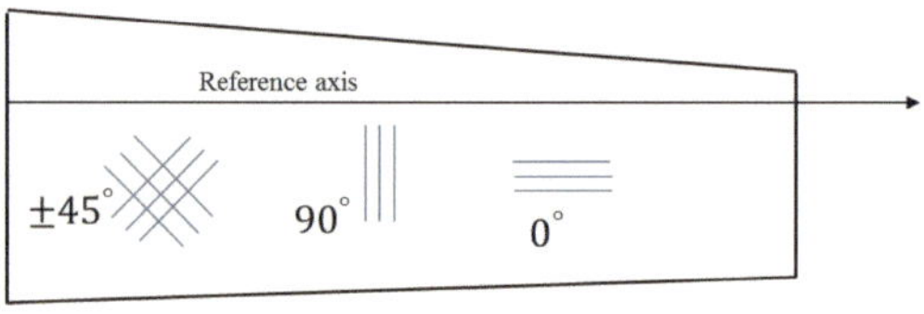

Figure 7.
Blade skin laminate configuration.

4. Blade construction and finite element analysis

The objective of wind turbine structure is to transfer and stand for wind turbine loads. Thus, it should be stiff enough to satisfy this objective. The structure weight is also important to be minimum as possible. A typical wind turbine structure consists of the skins, ribs, spar, and root or hub that connects between the blade and the wind turbine tower, as shown in **Figure 8**.

The ribs represent the aerodynamic profile shape for a blade. They distribute the aerodynamic loads and transform them to the main spar, in addition to maintaining the skin profile shape. The skins protect and cover the blade structure elements. Stringers stiffen the skins and connect between the structure of the skins and ribs. A spar represents the main structure element which carries the blade's main loads and transforms them to its root which in turn connects between the blade and the hub. A spar is usually consist of upper and lower flanges (caps) in addition to a shear web (**Figure 9**). The shear web performs high resistance to shear force in which the bending moment over the blade is transformed into in-plane shear forces that are carried by the shear web.

The geometric model available in **Figure 8** represents an equivalent model to the real wind turbine. This model usually is not suitable for finite element (FE) analysis. However, in finite element analysis, we use an equivalent model with some approximations that do not affect the accuracy of the analysis, but these approximations increase the speed and efficiency of the FE process. A geometric model that

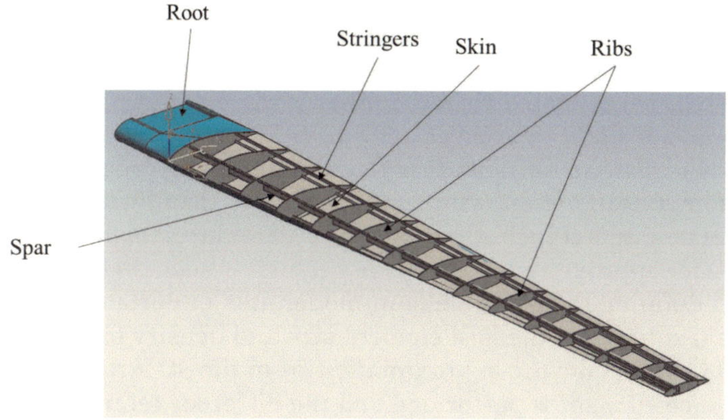

Figure 8.
A 6-m-diameter typical blade structure.

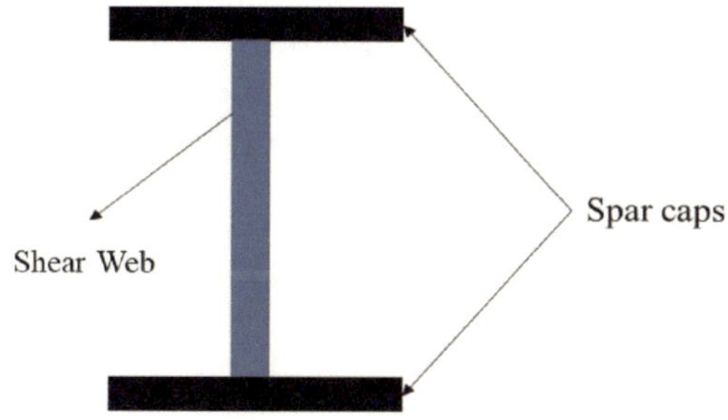

Figure 9.
A typical blade spar section (I—section).

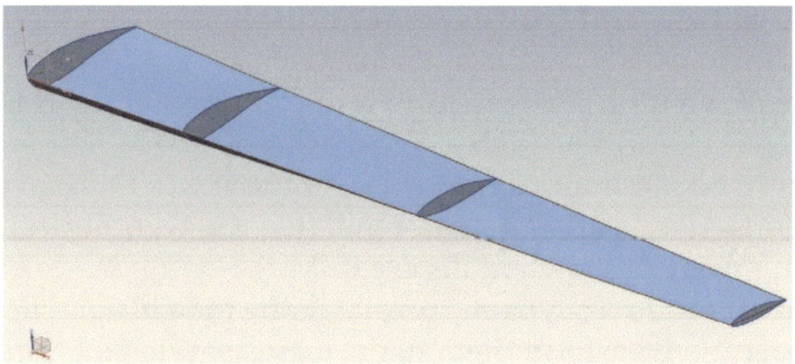

Figure 10.
Surface approximation for a blade rib and lower skin.

is physically a 3D model can be approximated into 2D or even 1D simple models. The blade skin and ribs, for instance, have a thickness dimension that is relatively smaller than the other blade dimensions; for this reason the skin and ribs can be approximated into 2D plates in finite element analysis. Geometrically, the blade skin and ribs are modeled as surfaces. Another example is the blade stringers. They have cross-section dimensions that are relatively smaller than the length dimension. Thus, stringers are usually modeled as 1D beams in the finite element analysis. These beams are geometrically represented by lines. These modeling concepts save significant time and effort in structural analysis and design process. An example of a surface approximation for a 6-m-diameter blade ribs and lower skin is shown in **Figure 10**.

In **Figure 10** the blade ribs and skin are represented by surfaces with zero thickness in which we assume there is no change either in stress or strain through their thickness. This approximation is valid as long as the thicknesses of the ribs and skin are relatively smaller than the other surface dimensions. Based on this approximation, a blade is meshed as a 3D model, but it is modeled numerically in FE using quadrilateral element. It is found that higher-order elements such as nine-node element can obtain results more accurate than linear quadrilateral elements in elastic and aeroelastic analyses [8]. So, the element selection is also important. After approximating the blade geometry, selecting the proper element, and defining the finite element model, it is important to perform a convergence test to select the best element size and density for the blade model [9]. **Figure 11** shows how the approximation from the 3D structure model to the 2D reference element is performed and the different reference coordinates that are used.

Figure 12 shows the geometric model for 10 MW wind turbine blade [11]. The blade has two spars, front and rear spars, in addition to 38 ribs.

The finite element process starts by approximating the physical model as we discussed, then defining the material properties, and selecting a suitable element. After that start the mesh process which transforms the continuous geometric model into discrete elements and nodes. Finally, the model boundary conditions are defined, and the model is solved. The objective of the finite element analysis is to solve the general equilibrium equation [12]

$$\sigma_{ij,j} + f_i = \rho u_{,tt} \tag{11}$$

in which $\boldsymbol{\sigma}$ is the Cauchy stress tensor, $\mathbf{f}$ is the force per unit volume, and $\mathbf{u}$ is the displacement. There are four applicable analyses based on this equilibrium equation:

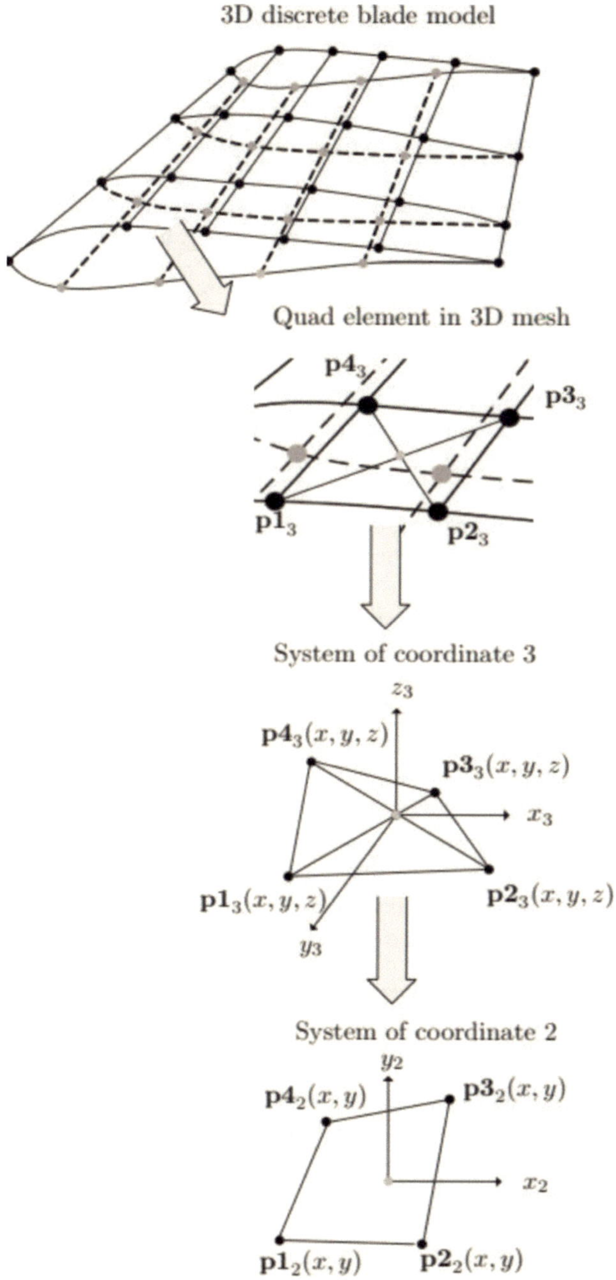

Figure 11.
From a 3D blade finite element mesh to a 2D quadrilateral element [10].

Figure 12.
A 140-m-diameter wind turbine blade geometric model [11].

4.1 Static structural analysis

We called this analysis one-way aeroelastic analysis in which the steady aerodynamic loads are defined and transformed to the structural model and then stresses, strains, and deformations are calculated. It is important in this analysis that the blade maximum deflection satisfy the minimum tower-to-tip clearance [11]. The maximum stresses and strains must be lower than the model allowable stresses and strains. The allowable stresses and strains are defined based on the material yield stresses or strains divided by a factor of safety.

4.2 Modal analysis

In modal analysis, we aim to calculate blade natural modes. Natural modes define the blade natural frequencies and their corresponding mode shapes at which the blade can fail if it excited at any of these modes. Therefore, a blade should be designed with natural frequencies higher than any surrounding frequency.

The modal analysis can be achieved by solving an eigenvalue problem based on the general equilibrium equation that has the form

$$(-\omega^2 \boldsymbol{M} + \boldsymbol{K})\boldsymbol{q} = \boldsymbol{0} \tag{12}$$

M and **K** represent the blade mass and stiffness matrices in the finite element model, respectively. **q** represents all the structural nodal degrees of freedom.

4.3 Dynamic analysis

In dynamic analysis, the unsteady aerodynamic loads are transformed into the wind turbine or blade structural model, and then the model is solved considering the time variation of both the loads and structural response. In dynamic analysis, the structure must be stable and safe from failure during its lifetime.

4.4 Aeroelastic analysis

A two-way aeroelastic analysis includes the calculation of blade aerodynamic loads and then transforms these loads into blade structure. After that the structure deformation is calculated and transformed back to the aerodynamic model to calculate new load distribution, and the process goes back and forth until a convergence point is determined. In the following section, we will discuss the wind turbine aeroelastic analysis process.

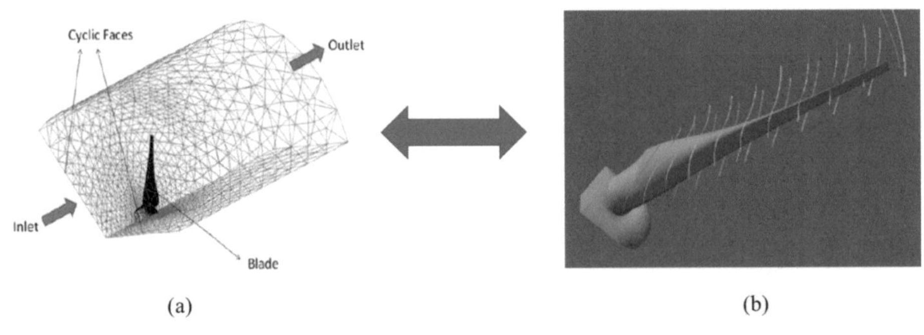

Figure 13.
(a) Aerodynamic loads calculated using CFD analysis. (b) The pressure distribution transformed to the blade structure [14].

One of the concerns in aeroelastic analysis is the connection between the aerodynamic model and the structural model. The aerodynamic model is usually made using CFD analysis, while the structural model is usually made using FEM. Different methods have been used to connect between the aerodynamic and structural models such as spline methods that are used in ANSYS software. It is found that the use of the finite element shape functions is more efficient in connecting between the aerodynamic and structural models, than the spline methods [13] (**Figure 13**).

5. Aeroelastic analysis of wind turbine blades

With the large increase in wind turbine size, the determination of aeroelastic instability of wind turbine blades becomes a must. Nowadays, wind turbines are designed with high speed, large scale, and light structure. All these specifications make them vulnerable to aeroelastic instabilities. There are two important things in aeroelastic analysis: (1) calculating the exact loads of a wind turbine blade and (2) determining the static (divergence speed) and dynamic (flutter speed) stability limits for a wind turbine.

5.1 Aerodynamic loads

In relatively large-scale wind turbines, the structure elastic nature allows it to deform significantly due to the applied aerodynamic loads. This deformation in turn affects the wind turbine thrust and output power. An example is the wind turbine blade shown in **Figure 14**. It is designed for a 1550-kN thrust; however, the wind turbine actual thrust is 1500 due to the aeroelastic effect.

As the wind turbine size becomes larger and larger, these effects become more significant. Therefore, for large-size wind turbines, the aeroelastic effects become more important and can affect significantly the actual wind turbine power. In other words, the wind turbine aeroelastic deformation can significantly decrease the wind turbine power and at the same time decrease the tip-to-tower distance which can lead to wind turbine failure.

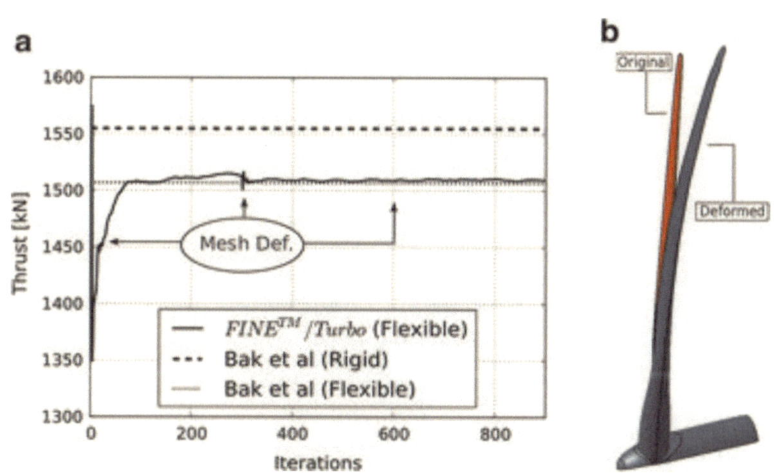

Figure 14.
Aeroelastic effect on the thrust of a 10 MW wind turbine blade [15].

5.2 Static and dynamic aeroelastic instability

The first time for an aeroelastic instability accident to appear was in Washington, USA, in 1940, as a result of Tacoma Narrows Bridge collapse. Then, the aeroelastic instability analyses become important for aerospace structure such as aircraft wings [16] and wind turbines. Recently, researchers start to investigate different aeroelastic phenomena in wind turbines. Aeroelastic instability analysis includes two aeroelastic phenomena: divergence and flutter. The divergence is

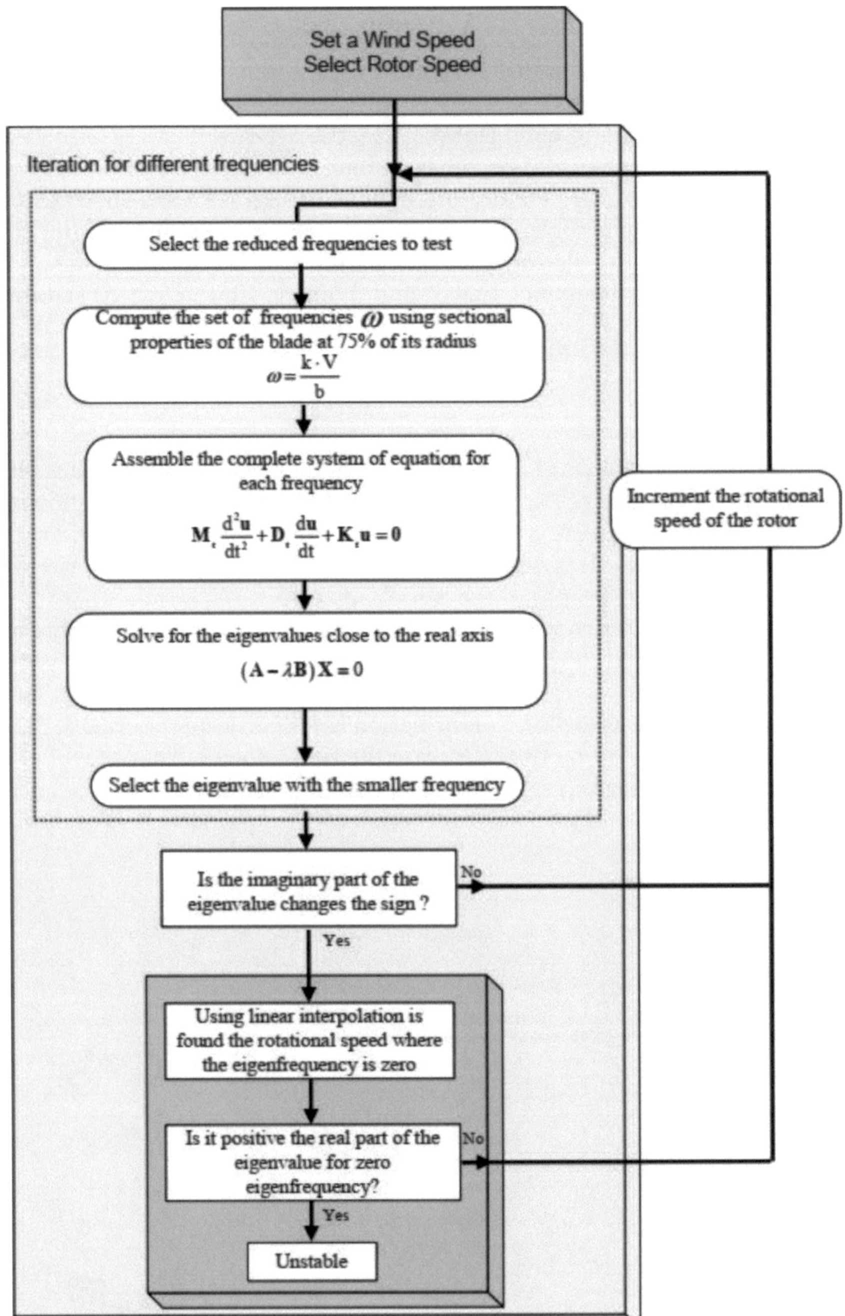

Figure 15.
A typical iterative process for finding a wind turbine blade divergence speed [10].

known as a static aeroelastic instability due to the blade's large torsional displacement in response to static aerodynamic loads. It occurs at certain speed known as the divergence speed. On the other hand, flutter is a dynamic aeroelastic instability

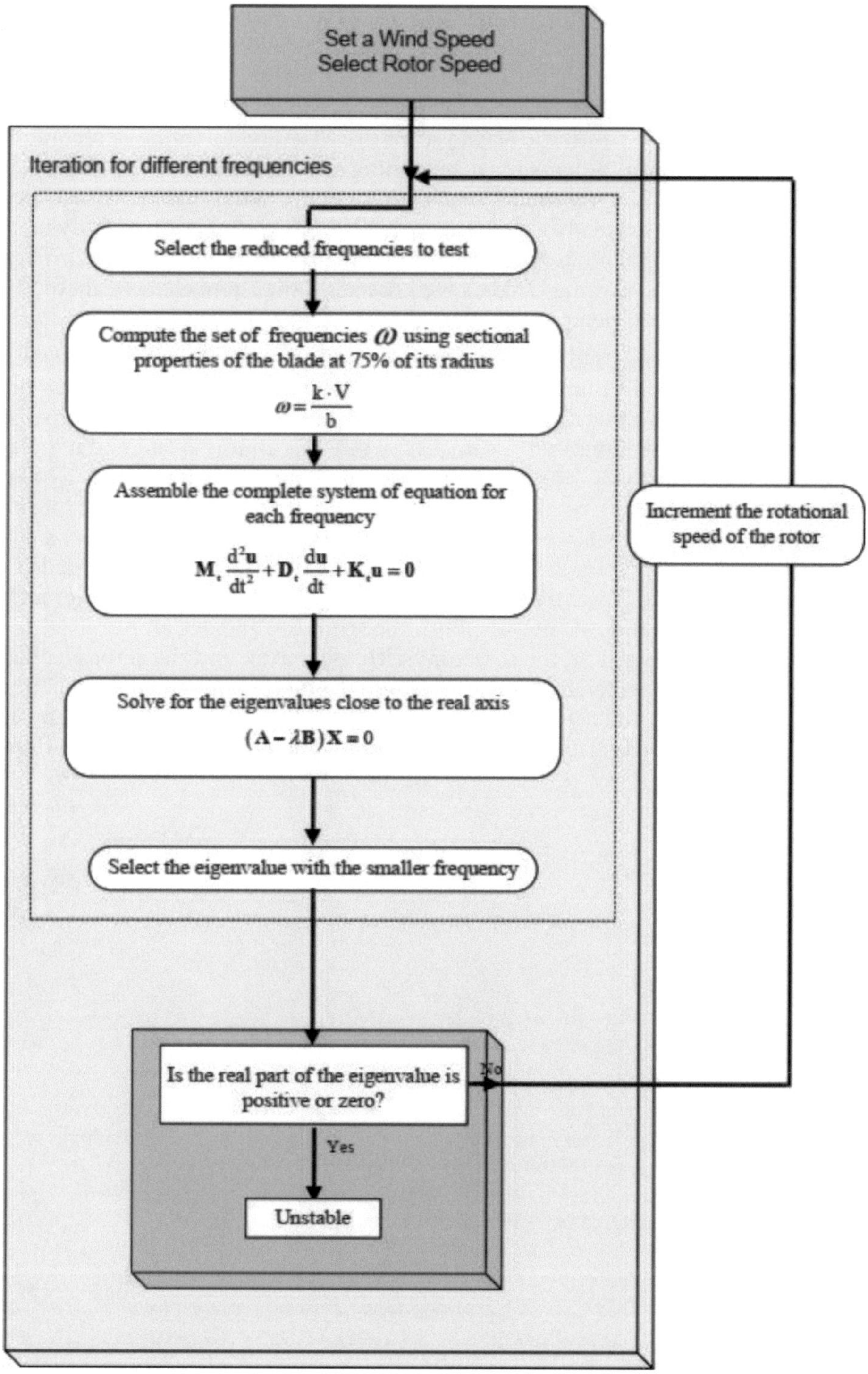

Figure 16.
A typical iterative process for finding a wind turbine blade flutter speed [7].

that occurs due to unstable vibrations that result from a coupling between twisting and bending modes at certain speed called the flutter speed.

No divergence or flutter failure is recorded for small-scale wind turbines (lower than 10 m diameter) [17] because small wind turbines have relatively stiff structure with low tip speed ratio. In such wind turbines, the flutter speed is found to be five times the wind turbine operating speed [17]. For MW wind turbines, the critical flutter speed becomes lower and lower, and subsequently the aeroelastic analysis becomes more important.

A wind turbine divergence speed is calculated based on steady aerodynamic loads in the interaction with static elastic structure based on an iterative technique. **Figure 15** shows a typical process for determining a wind turbine divergence speed based on the finite element method. On the other hand, a wind turbine flutter speed is calculated based on unsteady aerodynamic loads in the interaction with dynamic elastic structure based on an iterative technique. **Figure 16** shows a typical process for determining a wind turbine flutter speed based on the finite element method.

Notice that the finite element equation given in **Figures 15** and **16**, $\mathbf{M}_e \frac{d^2\mathbf{u}}{dt^2} + \mathbf{D}_e \frac{d\mathbf{u}}{dt} + \mathbf{K}_e\mathbf{u} = \mathbf{0}$, represents the general finite element equation derived from the structure equilibrium equation (Eq. (11)). The first term in the equation represents the structure inertia effects in which $\mathbf{M}_e$ is the element mass matrix. In divergence analysis the inertia term vanishes, while it is an important and mandatory term in flutter analysis. The second term defines the structure damping effect in which $\mathbf{D}_e$ is the damping term from both the structure and aerodynamic loads. Then structure damping can only calculate from the experiments. However, if no experiment data available for structure damping, the damping term is defined as a factor multiplied to the structure mass and stiffness matrices. The third term is the structure and aerodynamic stiffnesses. Both the structure and aerodynamic stiffnesses are calculated based on the structure stiffness matrix and the aerodynamic coefficient matrix, respectively. Once, the general finite element equation is defined based on the structure and aerodynamic analyses, it is transformed to an eigenvalue problem to determine the critical speeds for both divergence (**Figure 15**) and flutter (**Figure 16**).

Author details

Mohamed Abdou Mahran Kasem
Aerospace Engineering Department, Cairo University, Giza, Egypt

*Address all correspondence to: abdu_aerospace@eng1.cu.edu.eg

References

[1] Wind Energy Facts at a Glance | AWEA [Online]. Available: from: https://www.awea.org/wind-101/basics-of-wind-energy/wind-facts-at-a-glance [Accessed: 15 September 2019]

[2] WindEurope—The Voice of the Wind Energy Industry. WindEurope [Online]. Available from: https://windeurope.org/ [Accessed: 15 September 2019]

[3] Bangga G. Comparison of blade element method and CFD simulations of a 10 MW wind turbine. Fluids. 2018; **3**(4):73

[4] QBlade [Online]. Available from: http://www.q-blade.org/ [Accessed: 22 March 2017]

[5] Simulation Driven Product Development | ANSYS [Online]. Available from: http://www.ansys.com/ [Accessed: 07 February 2017]

[6] Ahmad T, Plee SL, Myers JP. FLUENT Theory Guide. ANSYS; 2011. p. 826. Available from: https://www.afs.enea.it/project/neptunius/docs/fluent/html/th/main_pre.htm

[7] Gasch R, Twele J, editors. Wind Power Plants. Berlin/Heidelberg: Springer; 2012

[8] Mahran M, ELsabbagh A, Negm H. A comparison between different finite elements for elastic and aero-elastic analyses. Journal of Advanced Research. 2017;**8**(6):635-648

[9] Mahran M, Negm H, Elsabbagh A. Aeroelastic Analysis of Plate Wings Using the Finite Element Method. Lab Lambert Academic Publishing; 2015. Available from: https://www.lap-publishing.com/catalog/details/store/gb/book/978-3-659-67969-8/aero-elastic-analysis-of-plate-wings-using-the-finite-element-method?locale=gb

[10] Capponi P. A finite element approach for aeroelastic instability prediction of wind turbines [Master]. [Thesis]. Delft, Netherlands: TU Delft; 2010

[11] Cox K, Echtermeyer A. Structural design and analysis of a 10MW wind turbine blade. Energy Procedia. 2012;**24**: 194-201

[12] Kasem MM, Dowell EH. A study of the natural modes of vibration and aeroelastic stability of a plate with a piezoelectric material. Smart Materials and Structures. 2018;**27**(7):075043

[13] Mahran M, Negm H, El-Sabbagh A. Aero-elastic characteristics of tapered plate wings. Finite Elements in Analysis and Design. 2015;**94**(Supplement C): 24-32

[14] Elqatary IHS. Experimental Characterization for Variable Pitch Horizontal Axis Wind Turbine. [Thesis]. Egypt: Cairo University; 2013

[15] Horcas SG, Debrabandere F, Tartinville B, Hirsch C, Coussement G. Hybrid mesh deformation tool for offshore wind turbines aeroelasticity prediction. In: CFD for Wind and Tidal Offshore Turbines. Cham: Springer; 2015. pp. 83-94

[16] Kasem MM, Negm H, El-Sabbagh A. Aeroelastic modeling of smart composite wings using geometric stiffness. Journal of Aerospace Engineering. 2019;**32**(2):04018143

[17] Vatne SR. Aeroelastic instability and flutter for a 10 MW wind turbine [Master]. [Thesis]. Trondheim: Norwegian University of Science and Technology; 2011

Chapter 6

Synthesis and Optimization of Wind Energy Conversion Devices

Janis Viba, Vitaly Beresnevich and Martins Irbe

Abstract

An approximate method for the analysis of interaction between wind flow and rigid flat blades is considered. The method allows synthesis and optimization of wind energy conversion systems without using space-time-programming procedures. By this method, the action of wind flow on the blade is subdivided on frontal pressure and vacuum (depression) on leeward side. The method was tested by computer simulation and experiments in wind tunnel. Examples of optimization tasks are solved in application to blades with simple shape. New wind energetic device with controlled orientation of flat blades to air flow is developed. Theoretical and experimental analysis of blade's interaction with airflow is performed. Aerodynamic coefficients for blade's drag and lifting forces are determined experimentally in wind tunnel. Optimization of system parameters is made. To increase the efficiency of energy transformation, it is proposed to change, by special law, the orientation of blade's working surface relative to airflow during rotation of the rotor. It is shown that the optimal angular rotation frequency ratio between rotor and blade is equal to two. Serviceability and main advantages of the proposed method are confirmed by experiments with physical model of airflow device.

Keywords: wind energy conversion, rigid blades, airflow, experimental aerodynamics, simulation, optimization

1. Introduction

Extracting energy from wind flow is one of the most important renewable energy sources. Operation principle of wind devices is based on air flow action on blades mounted on special wheel and further transformation of air flow kinetic energy into the mechanical energy of wheel rotation [1, 2]. In order to increase the efficiency of wind energy transformation, it is proposed to change orientation of blades to air flow during rotation of main rotor. Position of blade relative to air flow is considered as an optimal, if resulting aerodynamic force gives maximal torsional moment about longitudinal axis of the rotor. This chapter considers solution of the problem stated by the theoretical and experimental analysis of blade's interaction with air flow in different aerodynamic conditions.

Solid body interaction with air flow is studied in many works. Lift and drag coefficients for flat plate interacting with air flow in the range of Reynolds number from 10^4 to 10^6 were analyzed by Haibo et al. [3]. It was shown that at small angles of attack lift–drag relation is parabolic, but at higher angles of attack, it is a circle with radius of unity. Stationary regimes with constant vectors of relative velocities are well researched and mathematically described (for example, see [4]).

Nonstationary regimes of interaction between continuous air flow and rigid body, as shown in [5], mathematically are more complicated. Such kind of interaction can be analyzed by numerical simulation with space time programming methods [6].

In this chapter, a simplified method for the analysis of interaction between air flow and rigid body is proposed. Method is based on separate consideration of flow-body interaction on pressure and suction (vacuum) sides.

2. Simplified model of air flow interaction with flat plate

To simplify the analysis, optimization and synthesis of wind energetic devices, it is proposed to use approximated model of flow-plate interaction without considering viscous effects of air medium. For this purpose, air flow interaction with the plate is subdivided on two phases: one interaction is on the pressure side and the other one-on the vacuum side (**Figure 1**).

To analyze air flow-plate interaction, the theorem on change of momentum of mechanical system in differential form, is used [7]. In accordance with this theorem, on the side of pressure the following equation can be written:

$$dmV\cos\beta = dNdt, \quad (1)$$

where dm is an elemental mass of air flow; V is a velocity of air flow; β is the angle between air flow direction and normal to the plate surface $L1$ (**Figure 1**); dN is the impact force in the normal direction to the plate surface $L2$; and t is time.

Elemental mass dm of air flow can be mathematically described by the equation:

$$dm = \rho VB\cos\beta\, dt\, dL \quad (2)$$

where dL is elemental length of plate's surface; B is width of the plate (in the case of two-dimensional task B = const); and ρ is the air density.

By the integration of Eq. (1), extra pressure and force along x-axis can be determined. Mathematically, the pressure distribution can be described by the expressions:

$$p_{L1} = V^2\rho(\cos\beta)^2 \quad (3)$$

$$p_{L2} = V^2\rho(\sin\beta)^2 \quad (4)$$

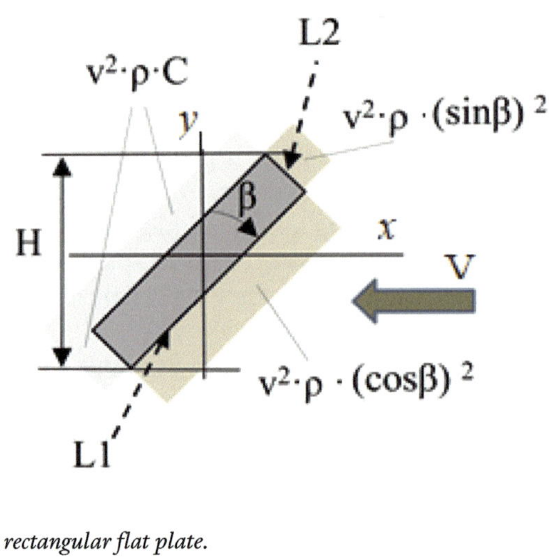

Figure 1.
Air flow interaction with rectangular flat plate.

Vacuum side of the plate is loaded by constant pressure, which can be determined by formula:

$$\Delta p_2 = +V^2 \rho \, C \tag{5}$$

where C is constant to be found experimentally or by computer simulation.

By this way, it is possible to find total force applied to the body in the air flow. For example, projection on x-axis of total force F_1 applied to the rectangular plate is described by the following equation:

$$F_{1x} = -HBV^2 \rho \left[C + \frac{(\cos\beta)^3 + d \cdot (\sin\beta)^3}{\cos\beta + d \cdot \sin\beta} \right], \tag{6}$$

where d is the ratio of edges $L2/L1$; H is section height of the plate in the direction perpendicular to the flow (**Figure 1**, $H = L1 \sin\beta + L2 \cos\beta$).

As follows from the results obtained, an approximate analytical method can be applied to solve air flow and blade interaction problems. Validation of the results of analytical calculations can be performed using computer simulation and experiments. Examples will be discussed in the following sections.

3. Theory and simulation of plane motion of flat plate

To determine coefficient C, numerical simulations in Solid Works for flat plate interaction with air flow were performed [8, 9]. Modeling results for a rectangular flat plate were compared with analytical formula (6). It was shown that approximate value of coefficient C is about 0.5. The estimation of accuracy of formula (6) for C = 0.5 in comparison with simulation results is presented in graphical form in **Figure 2**.

From the analysis of **Figure 2**, it can be concluded that interaction force fluctuates within certain limits, and value of fluctuation is dependent on angle β. At smaller angles, the value of fluctuations is also small, but with the increasing of β fluctuation of interaction force becomes greater. The relative difference is not very large, and its mean value is about 3.6%.

Therefore, engineering analysis by the proposed method does not require a step by step space-time calculations to find interaction coefficient C. A new nonstationary flow interaction formula was obtained, and this formula includes the object shape, state, and flow rate direction parameters. Coefficient C is taken as C = 0.5. The efficiency of the proposed method is illustrated by the analysis of flat plate interaction with air flow.

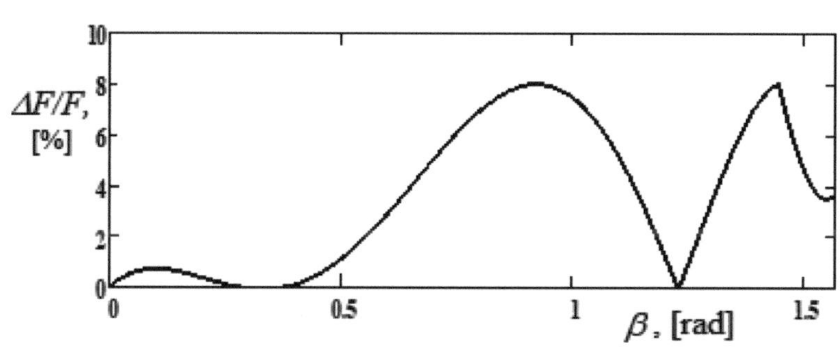

Figure 2.
The accuracy of formula (6) for C = 0.5 as function of angle β.

A two-dimensional (2D) model of translation motion of thin flat plate (thickness $d \sim 0$) in co-ordinate plane x–y is shown in **Figure 3**. The model includes linear elastic element with stiffness coefficient c and linear viscous damper with damping coefficient b.

In accordance with methods of classical mechanics [7], relative interaction velocity V_r must be taken into account:

$$V_r = V + v, \tag{7}$$

where V is velocity of air flow; and v is velocity of flat plate along x-axis.

In this case, differential equation of plate motion along x-axis can be written in the following form:

$$m\ddot{x} = -cx - b\dot{x} - A\rho\left\{\left[0.5 + (\cos\beta)^2\right]\cos\beta\right\}(V + \dot{x})^2\text{sign}(V + \dot{x}), \tag{8}$$

where $A = LB$ is a surface area of the plate; ρ is the air density; β is plate angle against air flow; and m is mass of the plate.

The renewable energy is generated due to the action of damping force $(-b\dot{x})$. Therefore, momentary power can be determined by formula

$$P = b(\dot{x})^2. \tag{9}$$

The average power P_a during time t is determined by integration of Eq. (9):

$$P_a = \frac{\int_0^t b(\dot{x})^2 dt}{t}. \tag{10}$$

By the analysis of Eq. (8), it can be concluded that five parameters can be used to control the efficiency of this system. These parameters are as follows: c, b, A, β, and V.

Mathematical simulation of Eq. (8) was performed with program MathCad assuming the following values of main system's parameters: $A = 0.04$ m^2 (length $L = 0.2$ m and width $B = 0.2$ m); $V_0 = 10$ m/s; $\rho = 1.25$ kg/m^3 (at temperature 10°C).

Results of simulation for control action by angle $\beta = \frac{\pi}{2.5}\sin(7t)$ are presented in **Figures 4** and **5**.

Average power P_a (**Figure 5**) is presented as percentage of maximal power, which can be achieved under the plate's velocity less than one third of flow velocity.

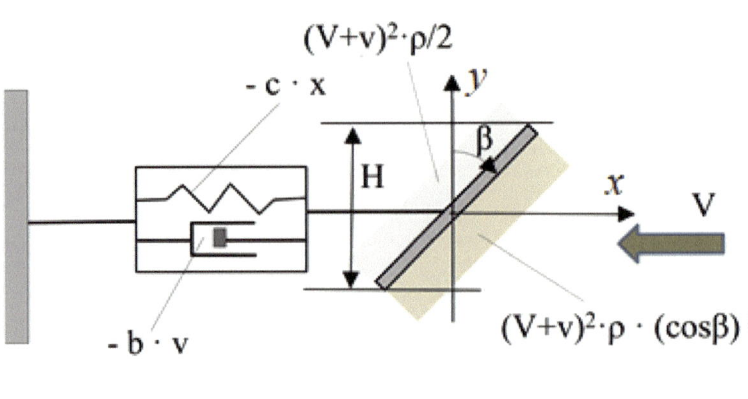

Figure 3.
Model of thin flat plate to obtain renewable energy from air flow.

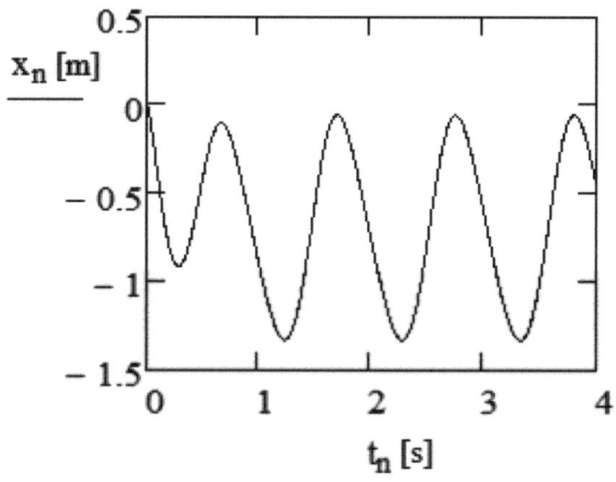

Figure 4.
Displacement x as function of time t.

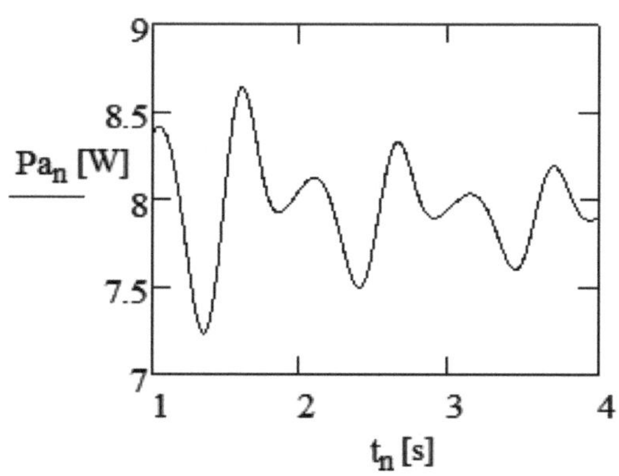

Figure 5.
Average power P_a of generator force $b\dot{x}$ during short transient process (t = 4 s).

Results of simulation for the case of control action on the system by harmonic variation of angle β (**Figures 4** and **5**) are as follows:

- stationary regime of motion occurs very quickly, practically within two to three cycles;
- it is possible to synthesize the optimal parameters of the system (for example, stiffness c, area A, frequency and amplitude of action, etc.), which would provide the maximum power within the given limitations;
- further increasing of the efficiency can be achieved by the use of more complex control variation of angle β (biharmonic, polyharmonic, etc.).

Results of simulation of plate's motion for the case of control action by velocity $V = V_0[2 - 0.5\sin(10t)]$ are presented in **Figures 6** and 7 (it is assumed that $V_0 = 10$ m/s).

As it is seen from the analysis of graphs presented (**Figures 6** and 7), almost stationary oscillatory regime with maximal average power P_a can be achieved after some cycles of transient process.

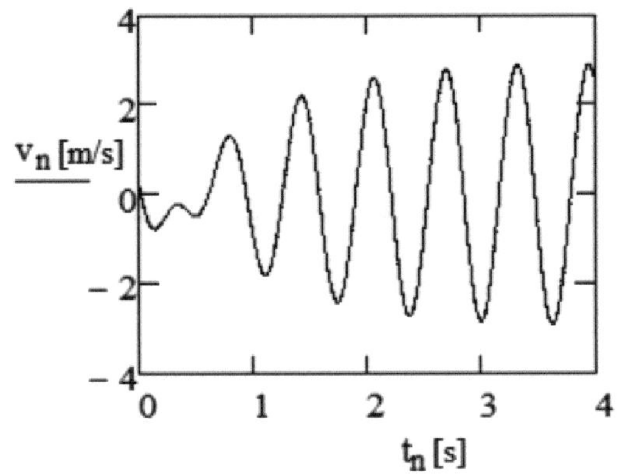

Figure 6.
Velocity v of plate's central point as function of time t.

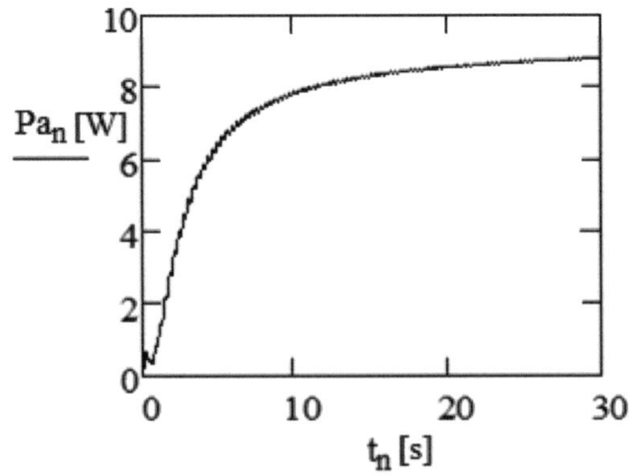

Figure 7.
Average power P_a of generator force $b\dot{x}$ during long transient process (t = 30 s)

Results of simulation for the case of control action on the system by harmonic variation of velocity V (**Figures 6** and 7) are as follows:

- a new opportunity to generate energy by the harmonic variation of flow rate is discovered;
- new variants of control action on the system by the flow rate variation using more complex laws (biharmonic, polyharmonic, etc.) are opened for the further research.

4. Parametric optimization of flat plate translational motion

The model of air flow interaction with flat plate is shown in **Figure 8**. In the given optimization problem, the power P generated due to action on the plate of air flow with velocity V is of interest. Parametric optimization aims to achieve maximum power P (criterion) providing plate's translational motion with constant speed u and in condition of braking with force Q (parameter).

In accordance with Eq. (6), the motion of a thin plate (d = 0) is described by the following differential equation:

$$m\ddot{x} = LB\rho(1+C)[V\sin(\gamma-\alpha) - \dot{x}\sin(\alpha)]^2 \cdot \text{sign}(V\sin(\gamma-\alpha) - \dot{x}\sin(\alpha))\sin(\alpha) - Q, \tag{11}$$

where m is mass of the plate; L and B are dimensions of the plate; α and γ are angles of flow and plate orientation against x axis.

The optimization task is solved under the condition of plate motion with a constant velocity $\dot{x} = u$. In this case, using Eq. (11), the braking power P of force Q can be expressed in the following form:

$$P = LB\rho(1 + C)[V\sin(\gamma - \alpha) - u\sin(\alpha)]^2 \cdot \text{sign}(V\sin(\gamma - \alpha) - u\sin(\alpha))u\sin(\alpha). \tag{12}$$

Parametric optimization problem was solved on the base of analysis of Eq. (12) with the aid of program MathCad. The aim of the optimization: to look for the parameters u, α, and γ that give the maximum power P at specified limits. The example of response surface for the criterion P as a function of plate angle α and velocity u is shown in **Figure 9**. In addition, **Figure 10** presents the section of this response surface corresponding to the constant velocity u = 8 m/s.

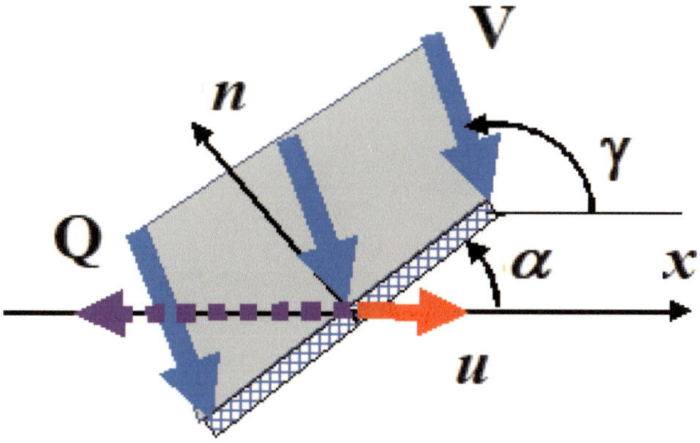

Figure 8.
Optimization model: V–air flow velocity; u–velocity of flat plate translation motion; and n–normal axis relative to plate.

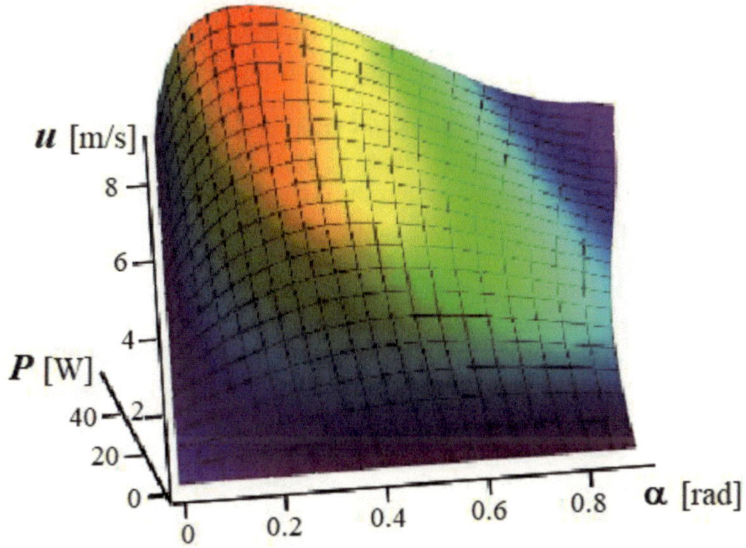

Figure 9.
Response surface for the optimization criterion P as a function of plate angle α and velocity u (for the case of L = 0.5 m; B = 0.5 m; ρ = 1.25 kg/m³; C = 0.5; V = 10 m/s; γ = π/2 rad).

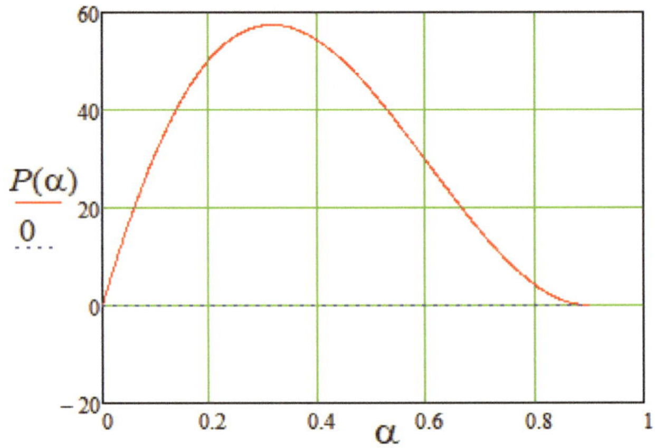

Figure 10.
Section of the response surface P(α, u) corresponding to the constant velocity u = 8 m/s.

The analysis of Eq. (12) shows that there is an extreme change of criterion P under the parameter u corresponding to the following condition:

$$u = \frac{V\sin(\gamma - \alpha)}{3\sin(\alpha)} \tag{13}$$

Under the velocity u corresponding to the condition (13) the maximal power P is achieved:

$$Pmax = \frac{4C_0V^3[sin(\gamma - \alpha)]^3}{27}, \tag{14}$$

where $C_0 = LB\rho(1 + C)$.

For the other initial data, the optimal parameters must be sought numerically.

5. Wind energy conversion device with controlled orientation of blades to air flow

Efficiency of energy conversion at great extent is dependent on blade's orientation relative to air flow. In majority of existing air flow devices, there is no possibility for special change of blade's orientation relative to air flow during rotation of main wheel, and therefore, position of blade can be optimal only in specific time instants [2].

Special variation of flat blade's turning angle during rotation of main wheel is realized in the device described in [10]. But interrelation between turning angles of blade and main wheel are not optimal in this device. Due to this, potential possibilities to increase efficiency of energy extraction are not fully realized.

Theoretical analysis of air flow interaction with rotating flat blade in different aerodynamic conditions is considered in [11, 12]. It is shown that position of blade is optimal, if resulting aerodynamic force gives maximal torsion moment about longitudinal axis of main wheel.

This section deals with the determination of optimal control law for variation of blade's turning angle during rotation of central wheel.

5.1 Analysis of air flow interaction with flat blade

The considered model of air flow device is shown in **Figure 11**. Flat blades 2 are hinged to the rotor 1, besides longitudinal axes O_1 and O_2 of rotor and blades are mutually parallel. Position of blade 2 relative to air flow is given by angle α, but rotation of rotor 1 is evaluated by angle φ. In order to increase the efficiency of wind energy transformation, it is necessary to find optimal relations between angles φ and α during operation of the system.

In accordance with aerodynamics theory [13], flat blade 2 placed in air flow is subjected to action of aerodynamic force R (**Figure 11**). Force R can be resolved into two components: drag force F_x (acts along flow direction) and lift force F_y (acts in direction perpendicular to air flow). The following formulae are used to calculate these forces [14]:

$$F_x = 0.5C_xA\rho V^2;\ F_y = 0.5C_yA\rho V^2 \tag{15}$$

where C_x and C_y–dimensionless drag and lift aerodynamic coefficients; A–area of blade's working surface; ρ–density of air medium; and V–velocity of air flow.

Aerodynamic coefficients C_x and C_y are dependent on blade's geometry, its orientation relative to air flow and dimensionless Reynolds number [13]. Coefficients C_x and C_y for the considered flat blade were determined experimentally in wind tunnel ARMFIELD. Principle diagram of experimental setup is shown in **Figure 12**.

During experiments, angle α between air flow and blade's flat surface was varied from 0° (working surface of the blade is parallel to air flow) to 90° (working surface of the blade is perpendicular to air flow), while the air velocity varies from 5 to 20 m/s. Aerodynamic forces F_x and F_y for different angles α and for the given flow velocity V were measured with the aid of weight mechanism (**Figure 12**). After that, dimensionless aerodynamic coefficients C_x and C_y were calculated using Eq. (15).

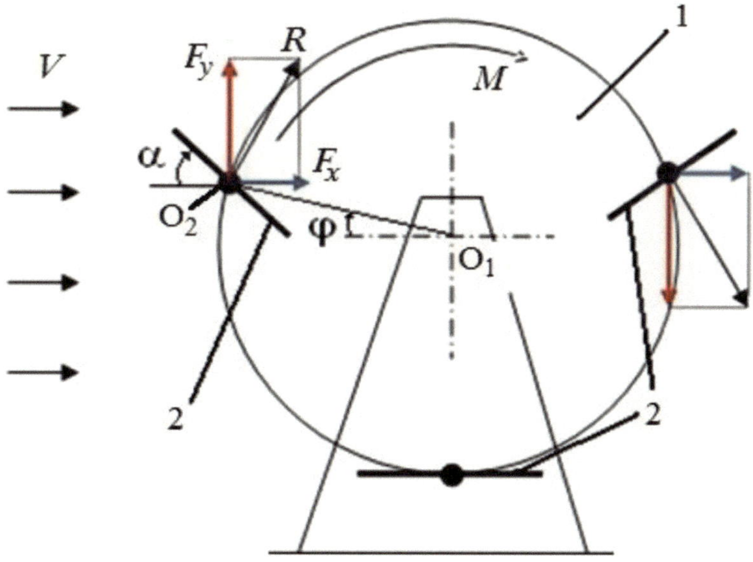

Figure 11.
Principle model of wind device: 1–central wheel; 2–flat blade.

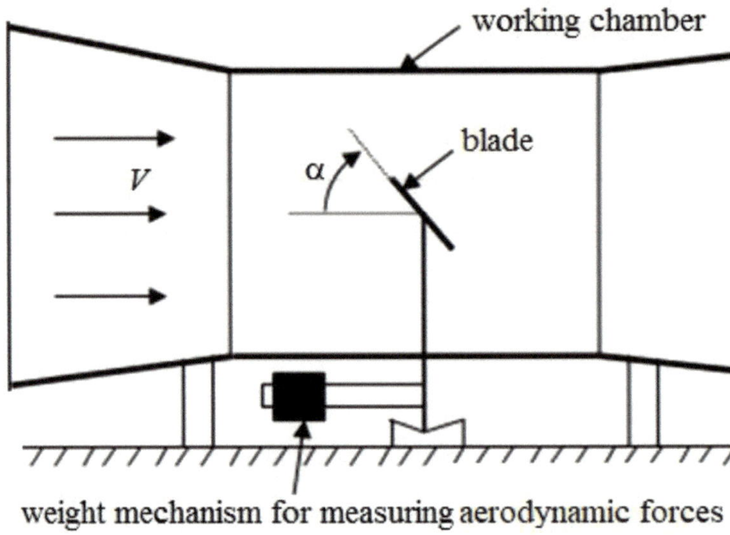

Figure 12.
Diagram of experimental setup.

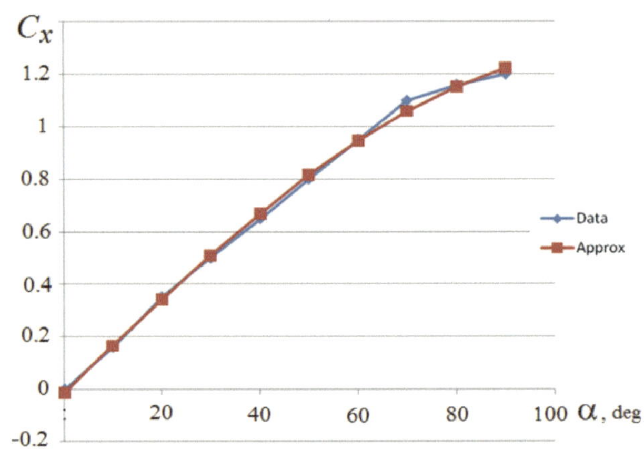

Figure 13.
Drag aerodynamic coefficient C_x as function of the incidence angle α.

On the base of mathematical processing of experimental results, coefficients C_x and C_y are plotted as functions of angle α (**Figures 13** and **14**, blue lines). To simplify the application of experimental data in engineering calculations, approximation of curves $C_x(\alpha)$ and $C_y(\alpha)$ is made (**Figures 13** and **14**, red lines) using program EXCEL.

Approximation functions $C_x(\alpha)$ and $C_y(\alpha)$ mathematically can be expressed with the following equations:

$$C_x(\alpha) = 1.31 \sin(0.0137\alpha) - 0.01 \tag{16}$$

$$C_y(\alpha) = -0.5094 \sin(0.00315\alpha - 0.28318)\alpha^{0.6143} \tag{17}$$

Coefficient C for resulting aerodynamic force R can be determined by the formula

$$C(\alpha) = \sqrt{[C_x(\alpha)]^2 + [C_y(\alpha)]^2} \tag{18}$$

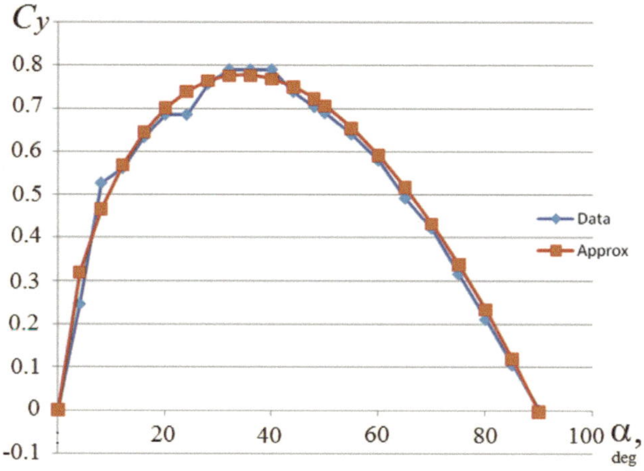

Figure 14.
Lift aerodynamic coefficient C_y as function of the incidence angle α.

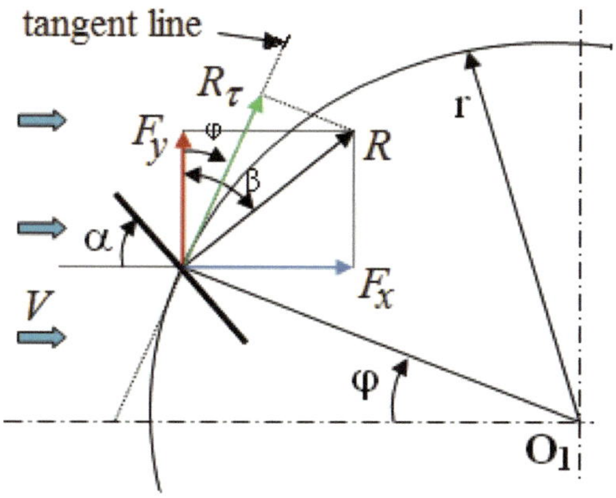

Figure 15.
Decomposition of aerodynamic force R on components.

For each value of rotor's turning angle φ, it is necessary to find the optimal incidence angle α. Angle α is considered as an optimal, if resulting aerodynamic force R gives maximal torsional moment M about longitudinal axis O_1 of the rotor (**Figure 15**). But moment M is determined by projection R_τ of force R on tangent line τ. Therefore, angle α is optimal, if projection R_τ becomes maximal. To satisfy this condition, it is necessary to maximize a projection C_τ of resulting aerodynamic coefficient C.

Projection of resulting coefficient C on tangent line can be determined by formula (**Figure 15**)

$$C_r(\alpha) = C\cos(\beta - \varphi), \tag{19}$$

where $\beta = \text{arctg}\left[C_x(\alpha)/C_y(\alpha)\right]$ the angles between vectors $\vec{R}$ and $\vec{F_y}$.

Optimal values of angle α were calculated with computer program MathCad using Eqs. (16)–(19). Calculations were made varying a turning angle φ of rotor from 0° till 360° with the step $\Delta\varphi$ = 5°. The results of calculations are presented in **Table 1**.

Using data of **Table 1**, a graph $\alpha = f(\varphi)$ is plotted (**Figure 16**). Curve $\alpha = f(\varphi)$ can be used in designing of wind devices to determine optimal positions of blade (angle α) for different possible values of rotor's turning angle φ.

φ[deg]	α[deg]	φ[deg]	α[deg]	φ[deg]	α[deg]	φ[deg]	α[deg]
0	35	90	90	180	145	270	180
5	37	95	91	185	148	275	180
10	39	100	92	190	150	280	181
15	41	105	93	195	152	285	182
20	43	110	99	200	154	290	183
25	46	115	105	205	156	295	185
30	49	120	110	210	158	300	188
35	52	125	115	215	160	305	190
40	55	130	119	220	163	310	193
45	58	135	122	225	165	315	195
50	61	140	126	230	167	320	197
55	65	145	129	235	170	325	200
60	70	150	132	240	172	330	202
65	75	155	134	245	175	335	204
70	81	160	137	250	177	340	206
75	87	165	139	255	178	345	208
80	88	170	141	260	179	350	210
85	89	175	143	265	180	355	212
90	90	180	145	270	180	360	215

Table 1.
Interconnection between rotor's turning angle φ and incidence angle α.

5.2 Prototype model of air flow device

To achieve maximal efficiency of wind energy transformation, it is necessary to take into account earlier determined optimal relations between turning angles φ and α (**Figure 16**). But exact curve $\alpha = f(\varphi)$ is a nonlinear one, therefore its practical realization in wind devices is very difficult.

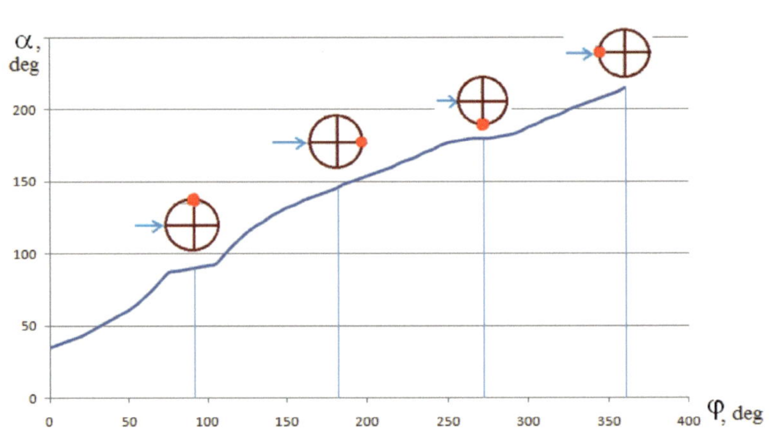

Figure 16.
The relationship between the optimal incidence angle α and rotor's turning angle φ.

To simplify practical realization of optimal angles α in real mechanisms, it is proposed to use linear approximation of exact (experimental) curve $\alpha = f(\varphi)$. Exact and approximated curves $\alpha = f(\varphi)$ are compared graphically in **Figure 17**.

Mathematically the linear relation between angles φ and α can be described by the following equation:

$$\alpha = 45° + 0.5\varphi = \pi/4 + 0.5\varphi \tag{20}$$

As follows from Eq. (20), one revolution of blade takes place after two whole revolutions of rotor. Therefore, during one cycle both opposite working surfaces of each blade take up a running air flow in turn.

Calculation error on application of approximate linear function (20) has been evaluated. For this purpose, values of angle α determined by formula (20) were inserted into the earlier used mathematical model (16)–(19). It was shown that in the most adverse case a relative error in determination of force R_τ does not exceed 10%. Herewith, as follows from **Figure 17**, approximate value of angle α during some stages of rotor motion is slightly smaller than optimal one, but in some other stages–a little more of it. Therefore, on average in one revolution, the difference between approximate and optimal values of angle α (or between generated powers) will be very small.

Besides, application of linear approximation (20) makes it possible to simplify practical realization of optimal operation regime in wind device. In accordance with Eq. (20), ratio of angular rotation frequencies between rotor and blades must be constant and equal to 2. Such constant frequency ratio can be realized using simple transmission (gear, belt or chain). Taking account of these considerations, it has been found expedient to use a linear approximation (20) in designing of wind device.

Kinematic diagram of the proposed air flow device is shown in **Figure 18**. Flat blade 2 is hinged to the rotor 1. Besides, longitudinal axes O_1 and O_2 of rotor and blade are mutually parallel and kinematically connected each other with belt transmission 3. Belt pulley 4 is attached to blade 2, but pulley 5–to rotor 1. And in accordance with Eq. (20), angular rotation frequency ratio between rotor 1 and blade 2 is taken as 2.

To follow possible changes of air flow direction, the proposed device is equipped with tail 6 fastened to pulley 5. Due to the change of air flow direction, tail 6 is turned till becomes parallel to air flow. Simultaneously pulley 5 and rotor 1 are also turned, as the result blade 2 takes optimal orientation to air flow. The proposed method for energy extraction from air flow and device for its realization are patented [15].

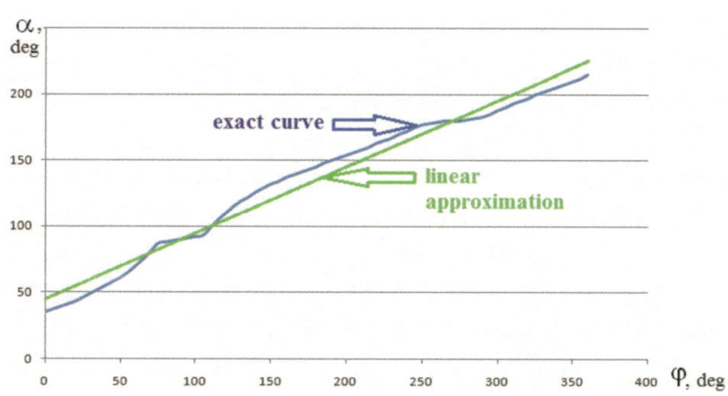

Figure 17.
Relationship between angles φ and α (experimental exact curve and its linear approximation).

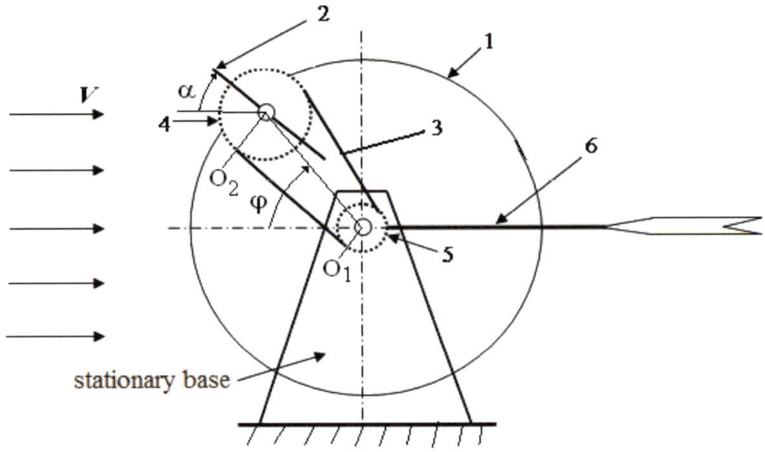

Figure 18.
Kinematic diagram of air flow device: 1–rotor; 2–flat blade; 3–belt transmission; 4 and 5–pulleys; 6–tail.

Figure 19.
Prototype model of wind energy conversion device.

In accordance with the proposed kinematic diagram, a prototype model of wind device is made (**Figure 19**).

Prototype model of wind device (**Figure 19**) contains four identical flat blades, which are kinematically connected with rotor by a toothed belt transmission. Ratio of angular rotation frequencies between rotor and blades is taken as 2; due to this the orientation of blades relative to air flow is changed in accordance with condition (20).

Experiments with prototype device are made in wind tunnel ARMFIELD (**Figure 20**). Stable and effective operation of the device is demonstrated for the range of flow velocities from 5 till 20 m/s.

Experimental investigations confirm the principal possibility to increase with the proposed method the angular velocity of central wheel and extracted power from air flow under the same wind velocity (in comparison with traditional air flow devices [14]).

Besides, experiments indicate on some shortcomings in operation of the prototype model shown in **Figure 19**. These shortcomings became especially evident with the increasing of number of blades. In such case the front blade (from the air flow side) can interfere for propagation of air flow to rearward blades. Air flow arriving to rearward blades can change its direction and intensity. Forming of air vortexes inside the space between blades has become possible. All these factors have negative influence on the efficiency of wind energy transformation.

In order to improve the efficiency of wind energy extraction, new design of air flow device is developed. Computer model of this wind device is shown in **Figure 21**.

Wind device contains three identical flat blades, which are fastened on axes hinged in central wheel. Blade's axes are kinematically connected with central rotor by a system of cylindrical gear transmission. In accordance with the condition (20), ratio of angular rotation frequencies between central wheel and blades is taken as 2. Besides, it is proposed to move blades relative each other along their rotation axes. Thanks to this, it became possible to make around each blade free space in radial direction for air flow access.

During rotation blades do not interfere each other (**Figure 21**). Each blade has its own open channel for passing of air flow. Therefore, application of this device will

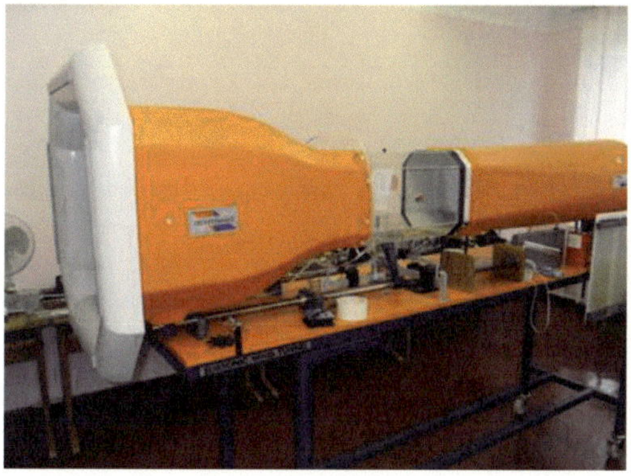

Figure 20.
Experiments in wind tunnel ARMFIELD.

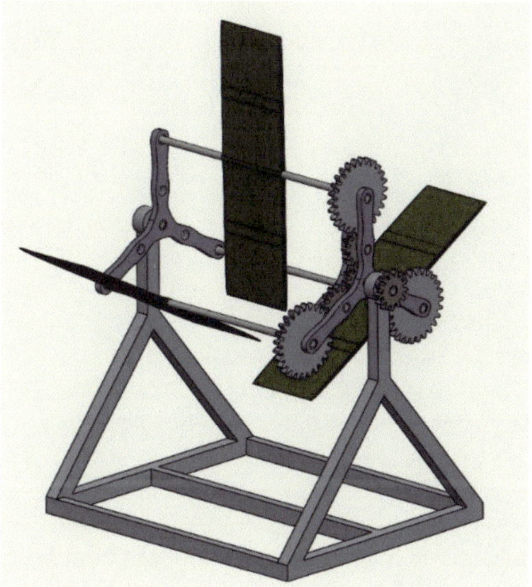

Figure 21.
New design of air flow device.

make it possible to increase the efficiency of wind energy transformation. More detailed quantitative analysis of operation of this device would be critical in the future.

6. Conclusions

New method for approximate analysis of air flow interaction with rigid body is developed. This method allows to solve problems of synthesis and optimization of wind devices in a simplified way, without using intensive and laborious space–time-programming procedures.

The aerodynamic drag and lift coefficients for flat blade are determined experimentally for different positions of blade relative to air flow.

Based on theoretical and experimental analysis, new designs of wind energy conversion devices with controllable orientation of flat blades to air flow are proposed.

A prototype model of the developed air flow device is given. Experiments confirm a serviceability of the device and efficiency of wind energy transformation.

Conflict of interest

The authors confirm that this section of monograph has no conflicts of interest.

Author details

Janis Viba, Vitaly Beresnevich* and Martins Irbe
Riga Technical University, Riga, Latvia

*Address all correspondence to: vitalijs.beresnevics@rtu.lv

References

[1] de Renzo DJ, editor. Wind Power: Recent Developments. Park Ridge, New Jersey: Noyes Data Corporation; 1979. 347p. ISBN: 9780815507598

[2] Dirba J, Levins N, Pugachevs V. Electromechanical Converters of Wind Energy (in Latvian). Riga: RTU Publishing House; 2006. 300p. ISBN: 9984320243

[3] Jiang H, Li Y, Cheng Z. Relations of lift and drag coefficients of flow around flat plate. Applied Mechanics and Materials. 2014;**518**:161-164. DOI: 10.4028/www.scientific.net/AMM.518.161

[4] Sears WR. Introduction to Theoretical Aerodynamics and Hydrodynamics. Reston: American Institute of Aeronautics and Astronautics; 2011. 150p. ISBN: 9781600867736

[5] Hossenfelder S. Is Space-Time Fluid? IAI News, Issue 59 [Internet]. 2017. Available from: https://iai.tv/articles/is-space-time-fluid-auid-897 [Accessed: 17 April 2019]

[6] Beal J, Viroli M. Space-Time Programming. Proceedings of the Royal Society A [Internet]. 2015. Available from: http://openmap.bbn.com/~jbeal/Publications/PTRSA2015-Space-Time-Programming-survey-preprint.pdf [Accessed: 15 April 2019]

[7] Meriam JL, Kraige LG, Bolton JN. Engineering Mechanics: Dynamics. 8th ed. New York: John Wiley& Sons; 2016. 736p. ISBN: 9781119044819 1119044812

[8] Tipans I, Viba J, Irbe M, Vutukuru SK. Analysis of non-stationary flow interaction with simple form objects. Agronomy Research Journal. 2019;**17**(S1):1227-1234. DOI: 10.15159/AR.19.083

[9] Vutukuru SK, Viba J, Tipans I, Viksne I, Irbe M. Analysis of flat plate vibrations by varying frontal area to the flow. In: Proceedings of 18th International Scientific Conference "Engineering for Rural Development". Jelgava: Latvia University of Life sciences and Technologies; 2019. pp. 1408-1414. DOI: 10.22616/ERDev2019.18.N147

[10] Elton LS. Windmill with Pivoting Blades. Patent US 7766602, USA, Int. Cl. F03D7/06/. Date of Patent: August 3, 2010

[11] Viba J, Eiduks M, Irbe M. Double pendulum vibration motion in fluid flow. In: Proceedings of International Conference "Engineering for Rural Development". Jelgava: Latvia University of Agriculture; 2015. pp. 434-439. ISSN: 1691-5976

[12] Viba J, Beresnevich V, Noskovs S, Irbe M. Investigation of rotating blade for energy extraction from fluid flow. In: Vibroengineering Procedia. Vol. 8. Kaunas: JVE International Ltd.; 2016. pp. 312-315. ISSN: 2345-0533

[13] Clancy LJ. Aerodynamics. New York: John Wiley & Sons; 1975. 610p. ISBN: 9780470158371

[14] Yanson RA. Wind Installations (in Russian). Moscow: Publishing House of Bauman MSTU; 2007. 36p. ISBN: 5-7038-2910-4

[15] Viba J, Noskovs S, Beresnevich V. Method for Control of Operation Condition of Wind Turbine and Device for Its Realization. Patent LV 15038, Republic of Latvia, Int. Cl. F03D7/06. Date of Patent: December 20, 2015

Section 2

Operational Monitoring, Reliability and Optimal Control of Wind Turbine Components

Chapter 7

Optimal Design and Operational Monitoring of Wind Turbine Blades

Francis Xavier Ochieng, Craig Matthew Hancock, Gethin Wyn Roberts and Julien Le Kernec

Abstract

The wind turbine blade is a critical component of any wind energy system. Its design, testing, and performance monitoring play a key role in power generation. With the increased use of composites and longer blades, a need to review existing monitoring sensors and use emergent novel ones is urgent among industry practitioners. In addition, an overview relating blade testing to Campbell diagrams and non-contact sensors have not been addressed as part of blade optimization. Based on design loads under IEC 61400-23 standards, the chapter explores various contact and non-contact sensors for design validation as well as their exploratory use in a three-tier structural health monitoring (SHM) framework for blade's operational performance monitoring. The chapter also includes a case study in the non-contact use of ground-based radar (GBR) in the optimal design of blades and real-time in-field monitoring using condition parameters. Lastly, the chapter addresses the lack of practical guidelines in the complementary use of GBR within a 3-tier SHM framework. Such use has the intent of building a cohesive understanding of GBR use for blade optimization and operational monitoring.

Keywords: wind turbine, blade structural monitoring, ground-based radar, sensors, Campbell diagram

1. Introduction

In-operation and design of load-carrying wind turbine (WT) blades, structural prediction of its vibratory characteristics are required in order to avoid resonance. Such characterization normally utilizes contact sensor-based forced-response and eigenvalue analysis. In detecting structural damage or obtaining validating condition parameters (CP) during design, the main practice is using either differential guided wave-based signal analysis or vibration-based damage detection (VBDD) [1]. Of the latter singular value decomposition (SVD) for structural damage detection has become widespread [1–3].

SVD works by comparing current sensor data to measurements taken from the healthy structure under varying environmental and operational conditions (EoC). Thus, the measurement from one sensor can be compared to prior acceptable or operating ranges (tolerance) of the static and unbalanced dynamic parameters of the structure.

In the case of WT, unbalanced parameters (also called condition parameters [CP]) are used. These are those parameters describing the resonant vibration due to unbalance of a vibrating or rotating object (i.e. divergence between the centre of geometry and centre of mass) [4]. Of the unbalanced parameters, [5–7] note that natural frequencies may comparatively be less prone to error than others unbalanced parameters like mode shapes and modal damping. They thus provide a ready application in using it for structural health monitoring (SHM) of the WT blades [8] both during design and operation.

To help in the acquisition of natural frequencies, the most widespread sensors used are gyroscopes and accelerometers [9–11]. These sensors, however, suffer from 3 key challenges: impractical to use on rotating WT blades, require laborious installation procedure, its results are vulnerable to EoC especially temperature, and lastly, the sensors are not easily portable. Other contact sensors like Fibre Bragg sensors are also widely used. Subsequently, there is a need to consider non-contact sensors like Laser or radar-based has become urgent.

2. Challenges in WT blades design optimization and operational monitoring

The singular goal of an optimization process or methodology is the exploration of the best possible solutions to a given problem. Implicit in this exploration is a continuum of solutions being considered against a set criterion. In WT modeling this would refer to maximizing or minimizing some function relative to some set options. Wherein the set would comprise of a range of choices existing at a time.

WT blades, unlike commercial aircraft blades, are comparatively more fatigue critical structures (see **Figure 1**). This means that their design is dictated not only by fatigue and ultimate strength but as also by aeroelastic considerations. For aircraft blades, their design is mainly dictated by fatigue and ultimate strength. The import of this being, as pointed out later, is that aircraft blades need only a few

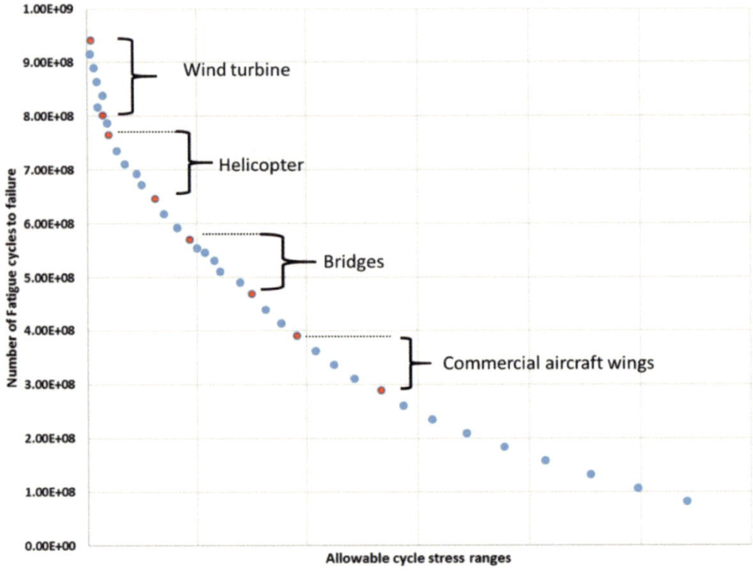

Figure 1.
S-N curves for various structures.

physical fatigue tests, unlike WT blades. This makes wind-turbine blade design, testing and certification expensive.

Usually, the approach for the WT type or component certification is physical testing using contact sensors to obtain metrics edge- and flap-wise deflection, mode shapes and modal frequencies for load cycles between 3×10^6 to 3×10^7, and to extrapolate the load cycles up to 109 for the lifetime of the wind turbine of 20–25 years using numerical simulations [12]. Generally, for a full-scale test, the allowable scatter between the actual tests and simulations is by a factor of 10 [12]. Two or more tests are thus needed to obtain a reliable conclusion.

The simulation is achieved by first undertaking a 10-min load simulation for a range of wind speeds. The results of each 10-min load analysis for each wind bin are then multiplied by 20 years, and the results for each load bin summed up to obtain a complete load distribution. Extreme loads are approached in a similar manner, but use made of a 50-year extreme 10-min average wind speeds with turbulence.

The challenges posed by these approaches are three-fold:

i. Rare events are not accounted for by the modeling framework [13]. This is bound to happen, since as the turbulence simulation is run, longer runs result in the tails of the stochastic inflow distribution of each average wind bin being filled with turbulence. Thus, a discrete averaged run multiplied by the 20 years lifetime may not capture this cumulative turbulence effect.

ii. Arising from the deficiency to capture the rare events, the statistics may not effectively estimate the peak load, which will occur during the wind turbine operating lifetime [13]. Currently, reliable estimates of such peak loads are not feasible due to lack of sufficient data. This is mainly due to the aerodynamic cyclic loading arising from wind shear and gravity making application of statistical modeling difficult.

iii. The IEC 61400-1 and IEC 61400-23 [14, 15] do provide specific design parameters that wind turbines designers can utilize. However, the standards do not specify

   a. The number of turbulence simulations is needed to establish the set probability and confidence levels for load predictions [13, 16].

   b. The load prediction processes, and material properties are decided based upon codified minimum safety factors that account for uncertainty. For homogenous materials, this would normally work. However, the use of composites presents a challenge due to its heterogeneous and visco-elasticity nature [16, 17].

   c. In addition to the design challenges, the EoCs of wind turbines are transient (non-periodic), consequently, the potential deflections during design and in real-life operating conditions are based on simulations in the time-domain [14–16]. This provides the minimum and maximum limits for the deflections that various sensors within the mast, and nacelle monitor. However, exact values are not easily determined for proper real-time system health monitoring.

Subsequently in **Figure 2**, the utilization of a non-contact sensor within a 3-tier SHM framework is demonstrated to significantly help in the design process, by providing real-time data on wind turbines in different operating environments. This it

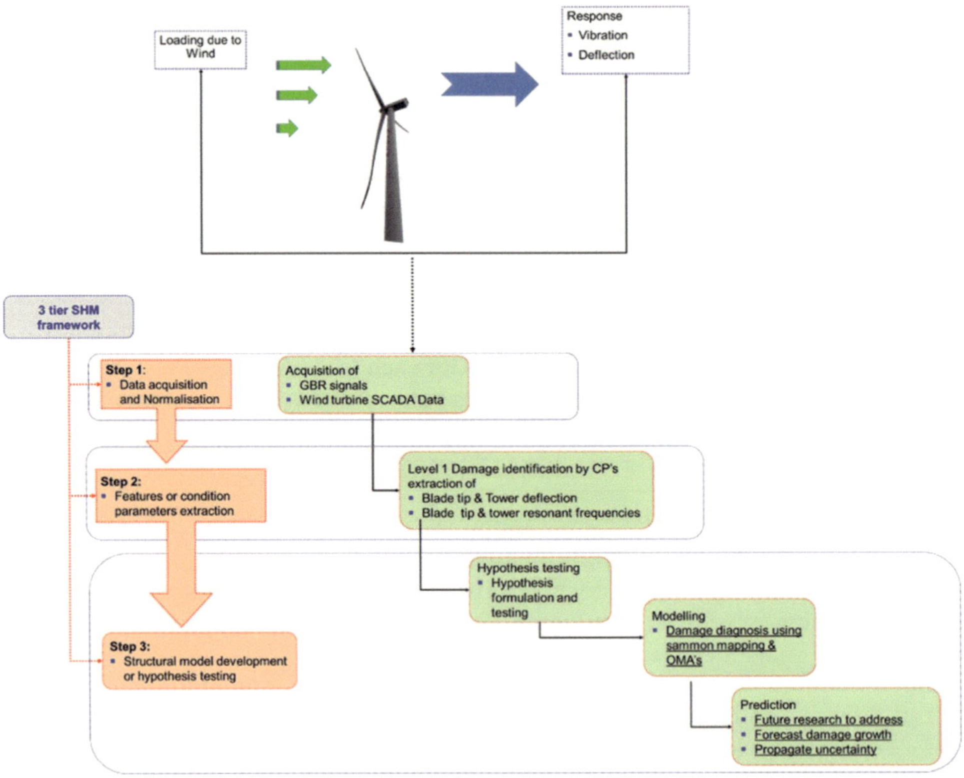

Figure 2.
GBR role in 3-tiered SHM of wind turbines.

achieves by enabling level 1 damage identification and prediction of future damage. A need thus exists to integrate such a non-contact sensor in not only the design process (**Figure 2**) but more importantly to help in the structural health monitoring (SHM) of WT using existing 3 or 4-tier SHM framework.

3. Tiered SHM framework for WT blades design optimisation and operational monitoring

Damage to WT structures can be identified much earlier with structural health monitoring (SHM) approaches. A recent study by [11], delineates two forms of SHM framework —a 3 or a 4-tier SHM framework.

The 4-tier SHM framework is based on 4 tiers/components [11], where:

- tier 1 is the determination if the damage is present,
- tier 2 locates the damage,
- tier 3 quantifies the severity of the damage, and
- tier 4 predicts the remaining service life of the system

Another SHM framework employs 3 sequential methodological steps viz.: (1) data normalization; (2) feature extraction using CP, and (3) health classification normally by way of hypothesis testing (HT) [11, 12].

In either SHM framework cases, data normalization provides a critical first step in health classifications since it compares the features of the structure in an unknown state (damaged or undamaged) to healthy features under the same environmental and operational conditions (EoCs).

Application of the 3-tier SHM framework has been done before [18] for a 3 kW small laboratory size WT. In that study [18] testing of the modularity of the 3-tier framework in SHM was undertaken. Data was collected from contact sensors placed in different parts of the WT. This data was then binned, analyzed and validated with receiver operating curves (RoC) to determine the damage classification capabilities of the framework. Residual-based CP's were utilized during feature extraction due to their sensitivity to both damage and EoC's.

Significant results from this study were that each tier can be considered as an independent (modular) and provide sufficient information for decision making on the structure's health. Secondly, that damage detection improved significantly if data clustering and binning were done before tier 2. However, the study was limited to the use of contact sensors for validation the 3 SHM framework and it used a comparatively smaller WT (3 kW).

A newer study [37] implemented the use of a non-contact Ground-Based-Radar (GBR) sensor within a 3-tier SHM framework. The study enabled the acquisition of frequencies and deflection condition parameters (unbalanced parameters as applied to SHM) with error margins of less than 10%. The application of the 3-tier SHM as applied in the study, for purposes of WT blades design and optimisation was as shown in **Figure 2**.

The role of SHM in dealing with the fatigue and emergent emphasis of aero-elasticity phenomenon like flutter is critical in integrating environmental and operational conditions (EoC) into SHM framework. The environmental conditions include: temperature, humidity, wind speeds, rainfall and irradiance whilst the operating conditions would include: cut-in and cut-out windspeeds, stall and pitch conditions, grid interconnection and whirling movements.

A review of various SHM approaches by [19] suggested that vibration-based damage detection (VBDD) methods provide the best SHM practices for beam-like structures. Other studies [20], however, suggest residual or differential imaged based signal analysis as being superior. For both approaches, damage occurrence (level 1 damage detection) [1] can be achieved. However [20], indicates that damage localization (level 2 damage detection) can be achieved only with the residual approach for real-life operating conditions. Such conditions were similar to those experienced by bridges or WT under operation.

4. Contact and non-contact sensors for WT

4.1 Contact sensors

The most common approach adopted to obtain the natural frequencies of structures is to place accelerometers at various locations on it and record the vibration responses from which the frequencies can be extracted. Accelerometers are commonly used due to their low cost and small mass which will not significantly affect the total mass of structures, hence vibration properties. However, the limitation of this approach is that the sensors need to be placed on the structures which may in some cases like those of rotating WT blades are neither accessible nor desirable.

In [37] a review of contact sensors is done within the context of wind turbine blade monitoring. This is summarized in **Figure 3**. Where the contact

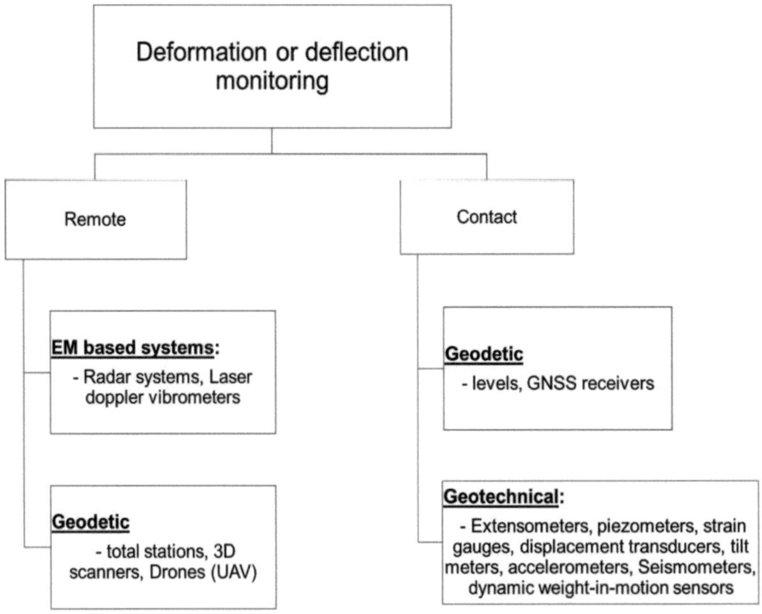

Figure 3.
Typology of sensors applicable to wind turbine blades.

and non-contact (remote) sensors are clustered in 3 main groups; Geodetic, Electromagnetic (EM) based and Geotechnical.

4.2 Non-contact sensors

While contact sensors like strain gauge sensors and accelerometers have been used in the monitoring of modal frequencies, the use of non-contact methods for beam testing has not been so widespread. **Table 1** depicts some of these non-contact sensors.

A number of non-contact methods like infrared thermography and photogrammetry have been demonstrated by [29, 33–35], for damage location under a 3-tier SHM. Other approaches like laser doppler vibrometer [26, 36] as a standalone or in conjunction with photogrammetry [6] have been employed in laboratory situations or for parked/non-rotating WT.

These methods do however face limitations particularly in the determination of modal frequencies, distributed strain, and deflection when the blades are in dynamic motion [37]. The main reasons for limitations are due to working principle employed, fast resolution of the EM, sound or light wave and environmental influences, which exacerbate the variations and errors in results.

4.3 Novel GBR for WT blades design optimisation and operational monitoring

Another novel approach has been the use of a quasi-monostatic ground-based real aperture radar (GBR). For this, a number of studies utilizing GBR for SHM have been done in the recent decade for beam-like structures (bridges and buildings) [38–40], towers [41–44] and for WT blades [11, 45, 46]. Specifically, the GBR provides a non-contact approach for design optimisation and operational monitoring of an operating WT [37].

As opposed to a monostatic radar using the same antenna alternately for transmission and reception, a quasi-monostatic radar has two antennas, one for

Sensor	Working principle	Limitations
Infrared thermography [21–25]	Utilizes infrared images to capture temperature occurring on damaged locations. Based on temperature increases of malfunctioning working components.	• Thermal images maybe unstable due to defocusing. • Can't acquire unbalanced parameters • useful for progressed or faults that have advanced.
Laser based systems [26-28]	Employs coherent radio waves to acquire modal parameters based on frequency shifts depicted in their interferograms.	• Input signal distortion and nonlinearity of the deflection mirror drive system • Cannot measure out-of-plane WT blade deflection due to speckle dropout errors (an optical phenomenon). • Cannot measure WT blades when in full rotation (>50m/s rotation). They limited to sensing up to 24.5m/s.
Photogrammetry systems [20, 29-32])	Employs either Digital Image correlation (DIC), target-less approaches, or 3 dimensional point tracking (3DPT)	• Requires optical reflectors or surface patterning mounted on WT blades. • Limited to low-frequency measurements. • High aeroelastic damping and dominant rotational harmonics influence results.

Table 1.
Non-contact sensors for WT blades monitoring.

transmitting and one for receiving. They are collocated that is the separation distance d_s is much less than the distance R between the radar antennas and the target ($d_s \ll R$) when compared to a bistatic radar [47]. Consequently, the equation to determine the maximum range (R) for monostatic radar is employed for quasi-monostatic radar.

When viewed from the GBR, the WT would consist of moving blades, almost stationary nacelle and a slightly moving tower. The nacelle is considered almost stationary since it rotates to enable to face the blades to the oncoming wind in addition to being stationary in situations of wind coming from a dominant direction. Generally, WT has large RCS in the order of 60 dBsqm (10^6 m^2) in the X-Band [48] and slightly less for the Ku band (~54 dBsqm), dependent on frequencies and blade aspect angle [49].

Even in the event of the blade rotation being low, the blade tip velocity will range between 50 and 150 m/s which is generally within the speed range of an aircraft. Hence providing a challenge to discern large WT from aircraft. This demerit, however, provides an advantage since it allows using Doppler frequency shift in the GBR backscattered signal, to distinguish between the tower, nacelle and blades.

It also allows the determination of the radial velocity (deflection velocity, v) of a target [50]. This is achieved by analysis of micro-doppler signatures, similar to the analysis of micro-doppler signatures of rotating helicopter blades [48]. Two important considerations are necessary in assessing such WT micro-doppler signatures:

- The radar cross-section (RCS) of a WT is much higher than that of a helicopter blade due to strong stationary reflections from the tower, nacelle and other ground clutter.

- The micro-doppler signatures may have doppler components of multiple bounces due to radar bouncing from blade to turbine tower to blades again before returning to the GBR.

The multi-bounce and stationary reflections may be best assessed using the design or expected operational parameters such as operational modal analysis (OMA's), in this case using Campbell diagrams.

The GBR acquires and processes the unbalanced/conditional parameters data in 5 key steps as shown in **Figure 4**. The data is acquired by radar, transformed into range profile using Fourier transform with possible windowing, thereafter deflection and modal frequencies CP's are obtained by phase extraction and power spectral density (PSD) respectively.

The maximum range R_{max} for the quasi-monostatic radar [47, 51] occurs when the received signal is equal to the minimum detectable signal S_{r_min}, and is found by Eq. (1).

$$R_{max} = \sqrt[4]{\left(\frac{P_t G_t G_r \lambda^2 \sigma}{(4\pi)^3 S_{r\_min}}\right)} \qquad (1)$$

S_{r_min} is the minimum detectable signal by the receiver antenna that would allow target detection and is expressed by $S_{r\_min} = k\,T_0 B F_n(SNR_1)$.

Where $k\,T_0 B$ is referred to as thermal noise from the ideal ohmic conductor, k is Boltzmann constant, T_0 standard temperature at 290 K, B receiver bandwidth, $F_n = (noise\ out\ of\ practical\ reciever)/(k\,T_0 B)$. For a signal to be detectable it has to be larger than the F_n by a factor called Signal to noise ratio (SNR_1). In addition, P_t is the transmitted power in Watts at the transmitter antenna, G_r and G_t are the antenna gains for the receiver and the transmitter respectively, while λ is the radar signal wavelength in m, and σ the radar cross-section area (RCS) in square metres.

The different time stamps of the return waves P_r, distinguishes them from each other allowing particular sections of the blade or mast to be identified in the

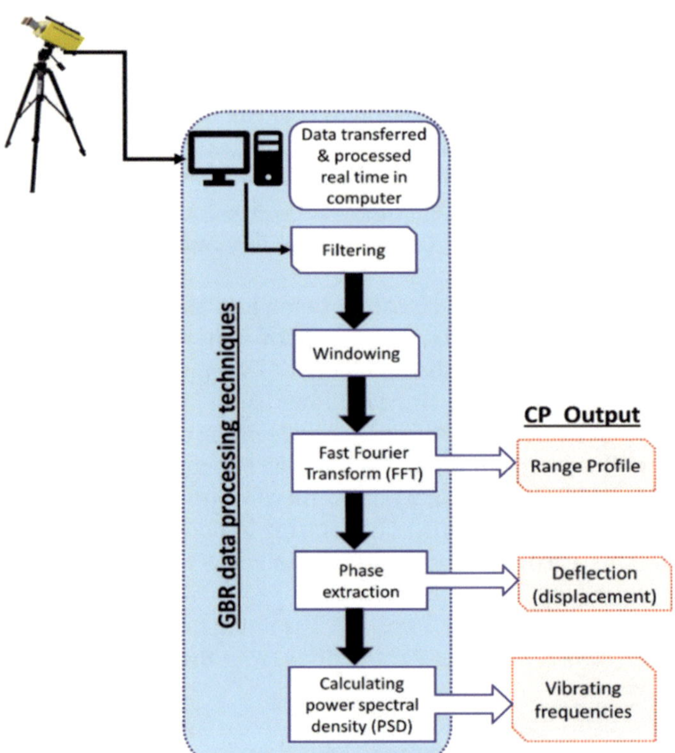

Figure 4.
GBR processing techniques.

corresponding time domain signal analysis. **Figure 5** demonstrates the setup for such signal acquisition; with pulse width τ in seconds and the inter-pulse period T_0 measured in seconds.

The product $P_t G_t$ is known as the effective radiated power (ERP), while $\sigma / 4\pi R^2$ is the fraction of the ERP intercepted and backscattered by the target.

A worst-case scenario is normally considered in order to know the maximum detection range R_{max}, which will occur when P_r is at its minimum [51] Eq. (2). P_r is inversely proportional to the fourth power of the range Eq. (2) [51].

$$\mathrm{P}_{rmin} \propto \frac{1}{R_{max}^4} \tag{2}$$

To determine the unambiguous range, Eq. (2) uses the inter-pulse period T_0 [51].

$$T_0 = \frac{2R_{amb}}{c} \equiv R_{amb} = \frac{cT_0}{2} \tag{3}$$

Where c is the speed of light (3×10^8 m/s) is the velocity of light and τ is the time taken by the radar to hit the target and return.

To acquire the modal frequency, which is twice the total Doppler frequency, Eq. (4) is used. Herein the total Doppler frequency (f_D) is the frequency shift obtained by the difference between the carrier frequency (f_O) and reflected signal (f'_O) [52, 53], in a blade tip movement away and back towards the radar.

$$f_D = f'_O - f_O \approx \frac{2v}{c} f_O = \frac{2v}{\lambda} \tag{4}$$

where λ is the wavelength corresponding to the frequency of the transmitted wave. Note that v (the radial velocity of the target along the LoS of the radar). Velocity is defined as positive when the object is moving away from the radar. v can be obtained from f_D and vice versa because they are proportional. The Doppler information can only be extracted by recovering the phase history of the signal over time and therefore requires the GBR receiver to have the phase information of its waves be constant (be coherent) [54].

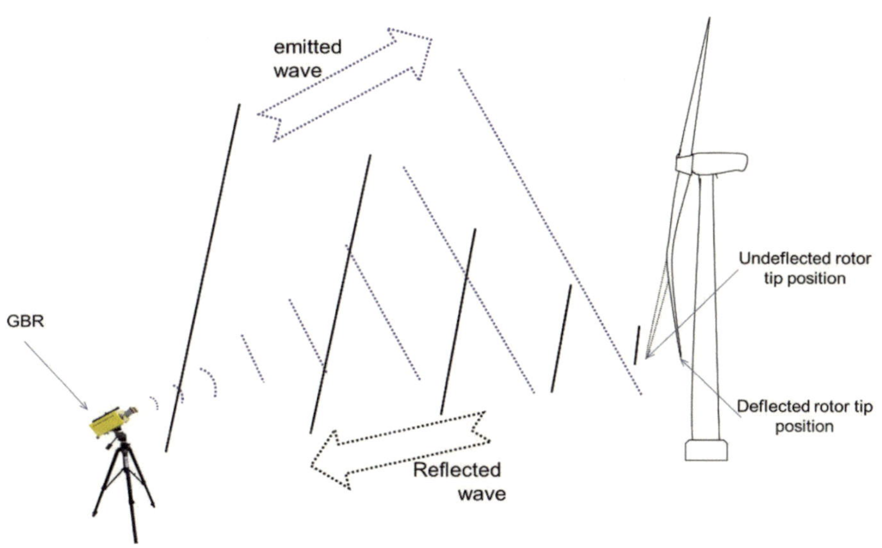

Figure 5.
GBR acquisition of unbalanced parameters using micro-doppler effects.

A 3-step process is utilized in radar target recognition that can be exploited for non-contact sensors application in a 3-tier SHM framework. The process entails

1. Acquire the Echo signal and analyze it using both SNR and RCS (tier 1 of the 3-tier SHM framework).

2. Feature extraction of target features from RCS sequences with known target category, then give a recognition criteria based on the relation between the target and its feature [55, 56] (tier 2 of the 3-tier SHM framework).

    - Feature extraction as a process aims to choose a subset of the original echo signal by the elimination of redundant information, yet extracting as much information as possible using as few features as possible [57]. Two approaches to features extraction are achieved by either

    - Extracting physical features from the time domain, such as extracting the cyclical nature of the RCS sequence [55] or

    - Extracting features from the transform domain (such as Fourier transform, wavelet transform, Merlin transform) [58].

3. Finally, recognize the damage or structure state by the recognition criteria (tier 3 of the 3-tier SHM framework).

The purpose of recognition criteria is to enable the identification of CP's and for this use can be made of principal component analysis or multidimensional scaling (MDS). MDS is a mostly a two-dimensional mapping or projection of data through the preservation of inter-point distances. It can either be a metric MDS like Sammon mapping or non-metric (neural networks, fuzzy networks, evidential and Bayesian approaches) [59, 60].

Of the four non-metric MDS methods—neural networks, fuzzy, evidential and Bayesian, the latter two provide the most relevance in terms of signal decomposition for damage recognition using recognition criteria. Evidential reasoning does not require prior knowledge of the probability distribution function. It is a method of fusing the different probability distribution functions given by different pieces of evidence. Thus give a recognition criterion based on the new probability distribution after fusing [57].

On the other hand, the Bayesian method requires the knowledge of the prior distribution. Then the minimum error rate or the minimum risk criteria can be given, and the target can be recognized by the criteria [57]. The Bayesian method in conjunction with non-contact sensors provides superior results in situations where no prior distribution existed either in the form of validated ground truth from contact sensors or in form of operational modal analysis techniques (OMA's).

The Campbell diagram is a form of OMA that is provided by the WT manufacturer for each wind turbine manufactured based on its design and potential operational parameters. Thus, it provides the apriori distribution of similar features required by the SHM framework.

5. Future trends and conclusions

While contact sensors have been widely employed in the design optimisation and monitoring of wind turbine blades, the use of non-contact sensors has not been fully highlighted. Specifically in the design optimisation and monitoring of blades, use of a 3-tier SHM framework and employing GBR are novel approaches. They offer new features and benefits in design and monitoring of WT blades.

With the advent of the fourth industrial revolution comprising of big data and internet of things, the GBR offers an opportunity to blend non-contact monitoring with improved design optimisation and monitoring of WT blades. One such technology is the use of GBR. However, future works in the deployment of GBR will need to focus on whirling movements of the WT nacelle and subsequent acquisition of condition parameters.

In conclusion this chapter has summarized the features and benefits as well as suggesting approaches and recommendations for future work, trends, and research. This is embedded in a conceptual framework that addresses the potential needs of WT blades trends in the future. It has further extended the complementary role and understanding of GBR in this role as a non-contact sensor, while proposing the integration of GBR as a non-contact sensor within the 3-tier SHM framework, to enable practitioners to undertake frequency based damage detection of WT blades. The main reasons for use of non-contact sensors is to address current challenges of installing contact sensors on operating/rotating WT, need for reduced SHM costs and lastly inappropriateness in use of contact sensors have had limited field and laboratory tests were undertaken on them.

Further, it has introduced the IEC 61400-23 standard for full structural monitoring of blades and relate it to sensors and Campbell diagram as an approach to optimization and operational monitoring of blades.

Author details

Francis Xavier Ochieng[1]*, Craig Matthew Hancock[2], Gethin Wyn Roberts[3] and Julien Le Kernec[4]

1 Jomo Kenyatta University of Agriculture and Technology, Nairobi, Kenya

2 University of Nottingham Ningbo China, China

3 The University of the Faroe Islands, Faroe Islands, UK

4 The University of Glasgow, UK

*Address all correspondence to: xavier@ieet.jkuat.ac.ke

References

[1] Loh C-H et al. Vibration-based damage assessment of structures using signal decomposition and two-dimensional visualization techniques. Structural Health Monitoring. 2019;**18**(4):1475921718765915

[2] Van Overschee P, De Moor B. Subspace Identification for Linear Systems: Theory—Implementation—Applications. Boston, London, Dordrecht: Kluwer Academic Publisher; 2012

[3] Liu YC, Loh CH, Ni YQ. Stochastic subspace identification for output-only modal analysis: Application to super high-rise tower under abnormal loading condition. Earthquake Engineering & Structural Dynamics. 2013;**42**(4):477-498

[4] De Queiroz M. An active identification method of rotor unbalance parameters. Journal of Vibration and Control. 2009;**15**(9):1365-1374

[5] Larsen GC, Hansen MH, Baumgart A, Carlén I. Modal analysis of wind turbine blades. Denmark: Forskningscenter Risoe; 2002

[6] Larsen GC et al. Effect of a damage to modal parameters of a wind turbine blade. In: EWSHM-7th European Workshop on Structural Health Monitoring. 2014

[7] Fernández-Sáez J et al. Unique determination of a single crack in a uniform simply supported beam in bending vibration. Journal of Sound and Vibration. 2016;**371**:94-109

[8] Bovsunovsky A, Surace C. Non-linearities in the vibrations of elastic structures with a closing crack: A state of the art review. Mechanical Systems and Signal Processing. 2015;**62**:129-148

[9] Sørensen BF, Lading L, Sendrup P, McGugan M, Debel CP, Kristensen OJD, et al. Fundamentals for remote structural health monitoring of wind turbine blades—a preproject. Roskilde, Denmark: Risø National Laboratory. Denmark: Forskningscenter Risoe; 2002

[10] Hameed Z et al. Condition monitoring and fault detection of wind turbines and related algorithms: A review. Renewable and Sustainable Energy Reviews. 2009;**13**(1):1-39

[11] Zhang S et al. UWB wind turbine blade deflection sensing for wind energy cost reduction. Sensors. 2015;**15**(8):19768-19782

[12] Pierik J, Dekker JW. European wind turbine standards II. ECN Solar & Wind Energy. 1998

[13] Veers P, Butterfield S. Extreme load estimation for wind turbines-issues and opportunities for improved practice. In: 20th 2001 ASME Wind Energy Symposium. 2001

[14] 61400-3, I. Wind Turbines–Part 3: Design Requirements for Offshore Wind Turbines. Tech. Rep., 2009

[15] Commission I.E.. International Standard IEC 61400-23 Wind Turbines–Part 23: Full-Scale Structural Testing of Rotor Blades. Geneva, Switzerland: IEC; 2014

[16] IEC. IEC 61400-1 Wind energy generation systems—Part 1: Design requirements. In: Onshore wind Turbines. IEC; 2016. p. 98

[17] Anant J. Design evaluation for IEC certification. In: 4th International Conference of Small Wind Association Testers (SWAT). Colorado, USA: Intertek; 2015

[18] Häckell MW et al. Three-tier modular structural health monitoring framework using environmental and operational condition clustering for data normalization: Validation on an operational wind turbine system. Proceedings of the IEEE. 2016;**104**(8):1632-1646

[19] Liu W et al. The structure healthy condition monitoring and fault diagnosis methods in wind turbines: A review. Renewable and Sustainable Energy Reviews. 2015;**44**:466-472

[20] Mitra M, Gopalakrishnan S. Guided wave based structural health monitoring: A review. Smart Materials and Structures. 2016;**25**(5):053001

[21] Tchakoua P et al. Wind turbine condition monitoring: State-of-the-art review, new trends, and future challenges. Energies. 2014;**7**(4):2595-2630

[22] Sanati H, Wood D, Sun Q. Condition monitoring of wind turbine blades using active and passive thermography. Applied Sciences. 2018;**8**(10):2004

[23] Beattie A, Rumsey M. Non-destructive evaluation of wind turbine blades using an infrared camera. In: 37th Aerospace Sciences Meeting and Exhibit. 1998

[24] Galleguillos C et al. Thermographic non-destructive inspection of wind turbine blades using unmanned aerial systems. Plastics, Rubber and Composites. 2015;**44**(3):98-103

[25] Meinlschmidt P, Aderhold J. Thermographic inspection of rotor blades. In: Proceedings of the 9th European Conference on NDT. 2006

[26] Stanbridge A, Ewins D. Modal testing using a scanning laser Doppler vibrometer. Mechanical Systems and Signal Processing. 1999;**13**(2):255-270

[27] Ozdemir. C, Inverse Synthetic Aperture Radar Imaging With MATLAB Algorithms. Hoboken, New Jersey: John Wiley & Sons Inc; 2012

[28] Prislan R, Svensek D. Laser Doppler Vibrometry and Modal Testing. Ljubljana: 2008. p. 17

[29] Poozesh P et al. Large-area photogrammetry based testing of wind turbine blades. Mechanical Systems and Signal Processing. 2017;**86**:98-115

[30] Lopez-Alba E et al. The use of charge-coupled device cameras for characterizing the mean deflected shape of an aerospace panel during broadband excitation. The Journal of Strain Analysis for Engineering Design. 2019;**54**(1):13-23

[31] Ozbek M et al. Feasibility of monitoring large wind turbines using photogrammetry. Energy. 2010;**35**(12):4802-4811

[32] Wang W, Li X, Chen A. A method of modal parameter identification for wind turbine blade based on binocular dynamic photogrammetry. Shock and Vibration. 2019;**2019**:10. Article ID: 7610930

[33] Liu W, Tang B, Jiang Y. Status and problems of wind turbine structural health monitoring techniques in China. Renewable Energy. 2010;**35**(7):1414-1418

[34] Yang R, He Y, Zhang H. Progress and trends in nondestructive testing and evaluation for wind turbine composite blade. Renewable and Sustainable Energy Reviews. 2016;**60**:1225-1250

[35] He Y et al. Volume or inside heating thermography using electromagnetic excitation for advanced composite materials. International Journal of Thermal Sciences. 2017;**111**:41-49

[36] Yang S, Allen MS. Output-only modal analysis using continuous-scan laser Doppler vibrometry and application to a 20 kW wind turbine. Mechanical Systems and Signal Processing. 2012;**31**:228-245

[37] Ochieng FX, Craig MH, Gethin WR, Julien LK. A review of ground-based radar as a noncontact sensor for structural health monitoring of in-field wind turbines blades. Wind Energy. 2018;**21**(12):1435-1449

[38] Saracin A et al. Investigations on the use of terrestrial radar interferometry for bridges monitoring. International Multidisciplinary Scientific GeoConference: SGEM: Surveying Geology & Mining Ecology Management. 2016;**2**:375-382

[39] Luzi G, Crosetto M. Building monitoring using a ground-based radar. In: Beer M, Kougioumtzoglou I, Patelli E, Au IK, editors. Encyclopedia of Earthquake Engineering. Berlin, Heidelberg: Springer; 2014

[40] Pieraccini M. Extensive measurement campaign using Interferometric radar. Journal of Performance of Constructed Facilities. 2016;**31**(3):04016113

[41] Muñoz-Ferreras J, Peng Z, Tang Y, Gómez-García R, Liang D, Li C. A step forward towards radar sensor networks for structural health monitoring of wind turbines. In: 2016 IEEE Radio and Wireless Symposium (RWS), Austin, TX. 2016. pp. 23-25. DOI: 10.1109/RWS.2016.7444353

[42] Metasensing. FastGBSAR case study—Structural Monitoring. 2016 [cited 2016 22.02.2016]; FastGBSAR case studies]. Available from: http://www.esands.com/pdf/Geotech/Metasensing/Real-Aperture-Radar.pdf

[43] Rödelsperger S, Meta A. MetaSensing's FastGBSAR: ground based radar for deformation monitoring. Proc. SPIE 9243, SAR Image Analysis, Modeling, and Techniques. 2014;**9243**(XIV):924318-924318-8

[44] Metasensing. FastGBSAR. 2016 [cited 2016 28th January 2016]; Product description of the GBSAR]. Available from: http://www.metasensing.com/wp/index.php/products/fastgbsar/

[45] Pieraccini M et al. In-service testing of wind turbine towers using a microwave sensor. Renewable Energy. 2008;**33**(1):13-21

[46] Pieraccini M. Monitoring of civil infrastructures by interferometric radar: A review. The Scientific World Journal. 2013;**2013**:8. Article ID: 786961

[47] Kolawole M. Radar systems, peak detection and tracking. 2003: Newnes

[48] Chen VC. The Micro-Doppler Effect in Radar. Boston, London: Artech House; 2011

[49] Jung J-H et al. Micro-Doppler analysis of Korean offshore wind turbine on the L-band radar. Progress In Electromagnetics Research. 2013;**143**:87-104

[50] Cheney M, Borden B. Imaging moving targets from scattered waves. Inverse Problems. 2008;**24**(3):035005

[51] Jenn D. Radar fundamentals. Department of Electrical & Computer Engineering. 2007;**93943**:831

[52] Luo Y, Yong-an C, Yu-xue S, Qun Z. Narrowband radar imaging and scaling for space targets. IEEE Geoscience and Remote Sensing Letters. 2017;**14**(6):946-950

[53] Levanon N, Mozeson E. Radar Signals. Hoboken, NJ, USA: John Wiley & Sons; 2004

[54] Mahafza BR. Radar Systems Analysis and Design

Using MATLAB. Florida: CRC Press; Taylor and Francis Group; 2013

[55] Wang T et al. Radar target recognition algorithm based on RCS observation sequence—Setvalued identification method. Journal of Systems Science and Complexity. 2016;**29**(3):573-588

[56] Lee JH et al. Performance analysis of radar target recognition using natural frequency: Frequency domain approach. Progress in Electromagnetics Research. 2012;**132**:30

[57] Kawalec A, Owczarek R. Radar emitter recognition using intrapulse data. In: Microwaves, Radar and Wireless Communications, 2004. MIKON-2004. 15th International Conference on. IEEE; 2004

[58] Franques VT, Kerr DA. Wavelet-based rotationally invariant target classification. In: Signal Processing, Sensor Fusion, and Target Recognition VI. International Society for Optics and Photonics; 1997

[59] Yin H. On multidimensional scaling and the embedding of self-organising maps. Neural Networks. 2008;**21**(2-3):160-169

[60] Tkac J, Spirko S, Boka L. Radar object recognition by wavelet transform and neural network. In: Microwaves, Radar and Wireless Communications. 2000. MIKON-2000. 13th International Conference on. IEEE; 2000

Chapter 8

The Use of Electrical Measurements of Wind Turbine Generators for Drive Train Condition Monitoring

Estefania Artigao, Andrés Honrubia-Escribano, Sergio Martín-Martínez and Emilio Gómez-Lázaro

Abstract

More modern and larger wind turbine (WT) generators are under continuous development. These exhibit more faults than smaller ones, which becomes critical offshore. Under this framework, operation and maintenance (O&M) is the key to improve reliability and availability of WTs, where condition-based maintenance (CBM) is currently seen as the preferred approach by the early detection and diagnosis of critical faults for WTs. The induction generator is one of the biggest contributors to failure rates and downtime of WTs, together with the gearbox and the drive train. In the present chapter, current signature analysis (CSA) will be introduced as a means for fault detection of WTs. CSA is a cost-effective and nonintrusive technique that can monitor both mechanical and electrical faults within the induction generator, as well as bearing- and gearbox-related faults. Different test cases of in-service wind turbine generators will be used to illustrate its usefulness.

Keywords: condition monitoring, current signature analysis, DFIG, wind turbine, operation and maintenance, predictive maintenance, drive train

1. Introduction

Wind energy is currently the most mature, cost competitive and cost-efficient source of renewable energy. This has been proven through the levelized cost of electricity (LCOE), which is a very common measure to define the cost of any sort of electricity generation. The historical evolution of onshore and offshore wind LCOE is shown in **Figure 1**. As can be seen, the LCOE has continuously decreased during the past few years both for onshore and offshore wind power generation. Offshore wind has experienced a bigger drop than onshore, although it still shows a significantly higher LCOE.

With over 590 GW of wind power installed globally, the market intelligence of the Global Wind Energy Council (GWEC) expects over a 55 GW per year global increment of new installations (combining both offshore and onshore wind) until 2023 [1]. In the present scenario, China leads the onshore sector, followed by the

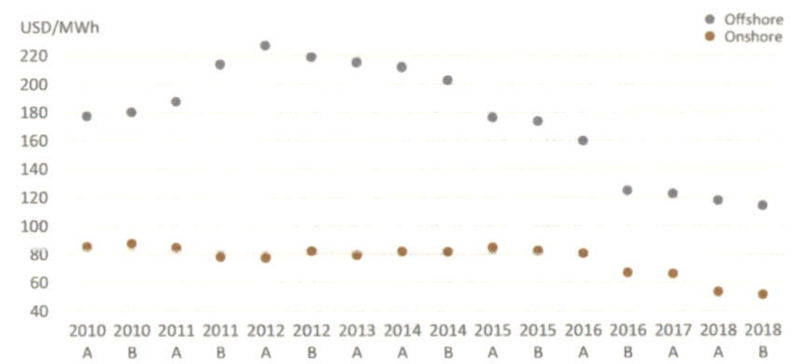

Figure 1.
Evolution of the LCOE for global onshore and offshore wind energy generation from 2010 to 2018.

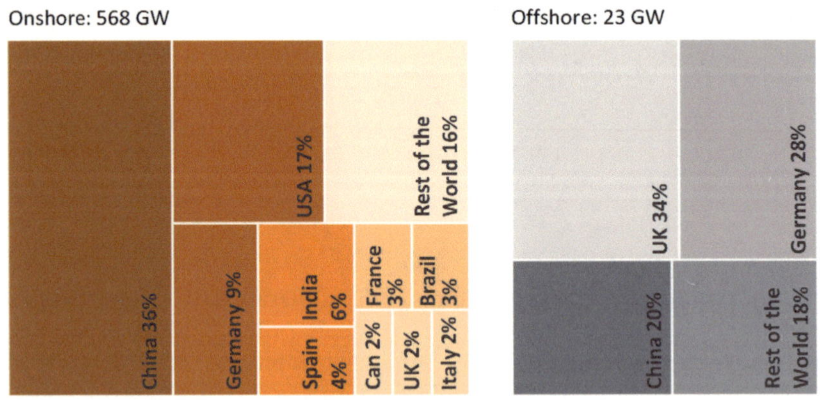

Figure 2.
Distribution of cumulative installed onshore and offshore wind energy capacity.

USA, Germany, India and Spain. The offshore sector is led by the UK, followed by Germany and China. The distribution of the top 10 countries with onshore wind power installed and the top 3 for offshore installations can be seen in **Figure 2**.

1.1 Reliability and availability

New sophisticated multi-MW wind turbines are constantly being developed [2–4], both for onshore and offshore applications [5]. Such rapid growth entails significant challenges that need attention from researchers and the industry [6]. In fact, more failures have been observed in large wind turbines than the small ones [7–9]. Current availability figures for onshore and offshore wind farms are found to be 97 and 85%, respectively [10, 11]. Accessibility issues and higher uncertainty are the reasons for the lower availability of offshore than the onshore.

Understanding which components are the most critical of a WT is the key to develop specific condition monitoring (CM) systems that will enhance O&M of wind farms, thus reliability and availability of WTs. To this end, this section presents the results published in the main failure rate and downtime surveys conducted in Europe, the USA and China. The major characteristics of these studies are summarised in **Table 1**, and the failure rate and downtime results presented in **Figure 3**.

As can be seen in **Figure 3**, both the electric and the control systems show the largest percentages for failure rates across the different studies (up to 32 and 42%, respectively) while hub and blades and gearbox for downtime (up to 38 and 56%, respectively). Medium percentages of failure rates and downtime are obtained for the generator, pitch and hydraulic systems. The braking system and drive train

Survey	Year	Location	WT number	WT capacity	On-/offshore
Acronym	**Published**			**[MW]**	
AWE [12]	2016	Europe	2270(av)	1–3	Onshore
CWEA [13]	2016	China	—	—	—
CARR [14]	2015	Europe	350	2–4	Offshore
B&W [15]	2014	UK	—	2.3	Onshore
CREW [16]	2013	USA	800–900	1.3–1.4	—
EZ [17]	2012	Netherlands	36	3	Offshore
RELI [18]	2011	Europe	350	>0.8	Onshore
VTT [19]	2010	Finland	72	0.2–2.3	Onshore
LWK [8]	2008	Germany	158–643	0.2–2	Onshore
WSDK [20]	2007	Denmark	851–2345	0.1–2.5	Onshore
WSD [20]	2007	Germany	1295–4285	0.1–2.5	Onshore
SW [7]	2007	Sweden	624(av)	0.5–1.5	Onshore
WMEP [9]	2006	Germany	>1500	—	Onshore

Table 1.
Characteristics of the main wind turbine reliability and availability surveys conducted in Europe, the USA and China.

present low failure rates but medium-to-high downtime. The sensors and structure categories present low values for both failure rates and downtime. Without considering the *others* category, the main contributors to failure rates across the different studies are hub & blades, electric, control and pitch systems. Similarly for the downtime, these are the gearbox, generator, mechanical brake and electric system assemblies.

If the main contributors both to failure rates and downtime are considered, then the control system, gearbox, electric system, generator and hub and blades could be considered the most critical components of wind turbines. Thus, special attention should be given to these components with regard to condition monitoring and O&M.

1.2 Operations and maintenance

One of the main leading costs of the wind farm's total expenditure belongs to O&M activities [12], representing between 25 and 35% for onshore and offshore wind farms, respectively, of their lifetime costs [21–23]. It is thus clear that reducing O&M costs would be beneficial for wind farm owners and operators.

Maintenance strategies can be typically classified as [21, 24] time-based (TBM), failure-based (FBM) and condition-based. TBM is formed by preventive tasks, also referred to as scheduled maintenance, and it is carried out regardless of the health state of the WT. FBM involves corrective actions, only performed after a critical failure has taken place, thus usually causing long downtime periods. CBM, on the other hand, is a predictive type of maintenance. CBM takes place only when there is a need but before a critical failure occurs, with the consequent saving in resources and downtime, thus achieving the optimum strategy [25, 26]. The different maintenance strategies are illustrated in **Figure 4**. Traditionally, a combination of TBM and FBM has been the preferred option among wind farm operators [17]. Lately, however, the trends are shifting towards CBM, which is performed through appropriate condition monitoring, recognised by [27, 28] as the key for optimised reliability and availability of wind turbines while achieving cost reduction in O&M [29, 30].

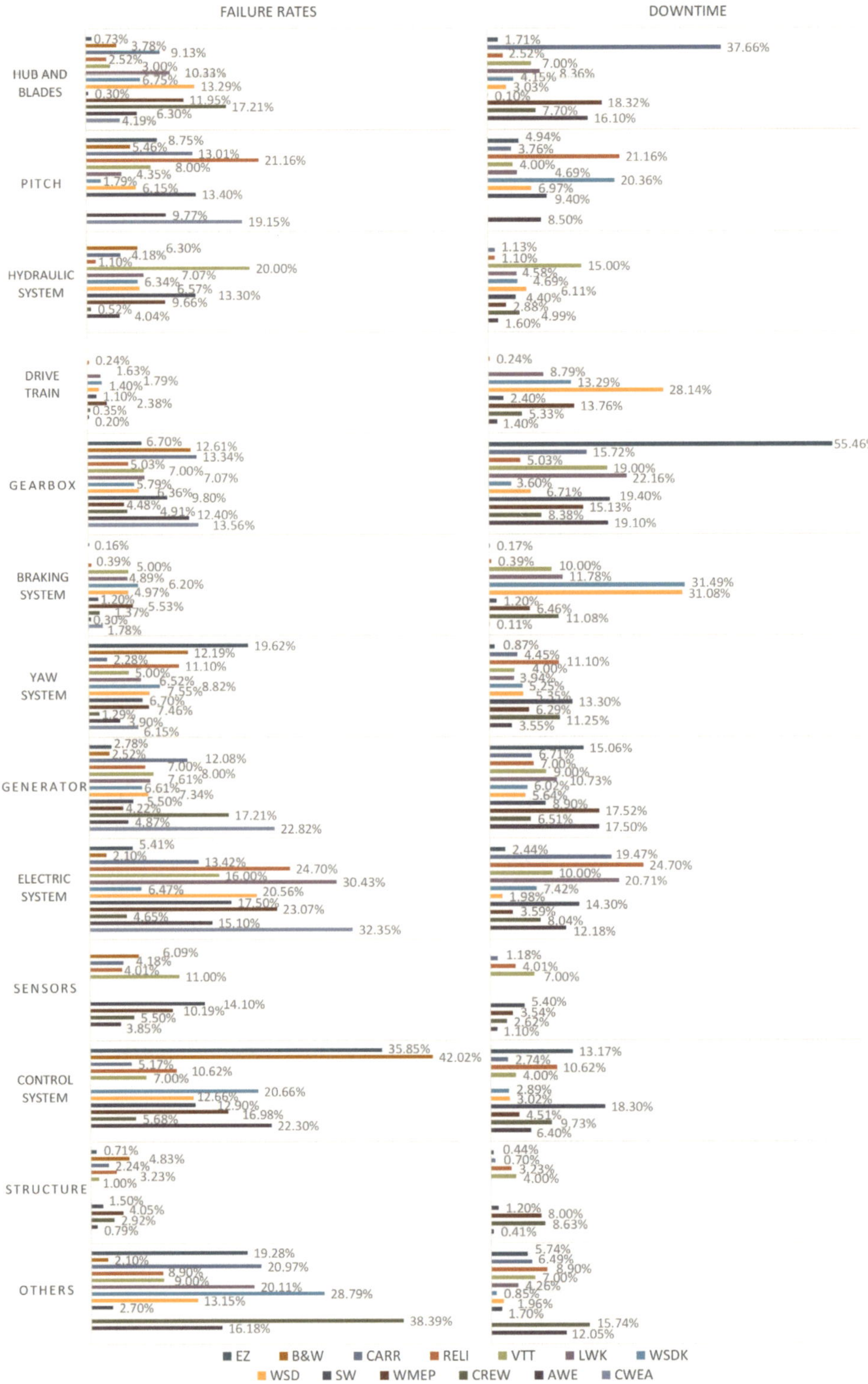

Figure 3.
Failure rates and downtime of several wind turbine reliability and availability studies spread in Europe, the USA and China.

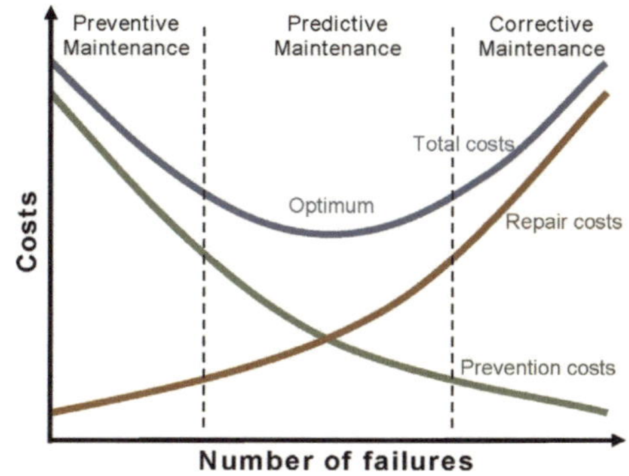

Figure 4.
Maintenance strategies.

1.3 Condition monitoring

CM consists of analysing different types of signals (such as temperature, vibrations, electrical measurements, etc.) with the aim of assessing the health of the component under study [24] towards calculating its remaining useful life (RUL) [31]. In this context, the implementation of CM systems must enable early-stage fault detection, so that CBM can be implemented, thus significantly reducing or even eliminating unscheduled (or unplanned) maintenance costs [32, 33]. Many CM and fault diagnosis reviews have been published in the last 15 years that include the latest CM methods [34, 35]. These can be classified per WT subsystem [36, 37] and differentiated between structural [38], mechanical [39], power electronics and control systems [40] or electrical parts [41–43].

Current commercial CM systems mainly target the WT drive train (particularly the gearbox) and are based on vibrations and oil analysis. The first ones to appear, and therefore the most common, were based on oil analysis. These systems, in addition to providing information on the status of the oil, are also capable to inform about the gears. They detect ferrous and non-ferrous particles, where the presence of ferrous particles is indicative of gear wear, while the presence of non-ferrous particles indicates oil degradation. Soon, a lot of research emerged based on vibration analysis, targeting the gearbox and bearings in the drive train. This type of analysis was adapted from other industries with rotating machinery. The latest commercial CM systems to appear were based on SCADA data. There are a large number of scientific publications utilising different parameters acquired by the SCADA system able to provide information of several components of the WT [44, 45]. Less common CM systems are based on acoustic emissions, strain gauges, thermographic analysis, etc.

On the other hand, with a non-negligible number of world-wide wind farms reaching the end of their expected 20-year lifetime [46], the field of lifetime extension (LTE) of wind turbines is gaining attention, where CM will also play a key role [47]. Different approaches are found for LTE of WTs such as practical inspections, field data analyses or analytical simulations. The authors in [48], for example, have identified data-driven approaches as the most cost-effective with large number of WTs.

1.4 Motivation

Data collection and analysis is not straightforward. There are many different techniques for signal processing and feature extraction, and, to date, no agreement has been established on which CM system is the most appropriate for specific components or assemblies. Furthermore, most of the techniques and methods developed for CM of WTs have only been applied on laboratory benches. In fact, with regard to the generator, despite being one of the most critical components in a WT, no commercial CM system has yet been developed that can provide exhaustive information about its health status.

The goal of the present chapter is to present a CM method based on electrical measurements that can detect different generator faults and has been implemented on in-service operating WTs. CM methods based on electrical measurements can be performed using current, voltage, instantaneous power and flux analyses [49]. Of these, stator current spectral analysis (Current Signature Analysis, CSA) is recognised as the preferred option [50–52], thus chosen for this book chapter.

2. The method: current signature analysis

CSA is a nonintrusive method capable of identifying electrical and mechanical faults in a cost-effective manner [53–55]. It is based on the principle that different faults have different effects on the current spectra. It can be implemented on doubly-fed induction generators (DFIGs), which are the most common technology on variable speed WTs.

As reported in [56], bearing-related faults are the most common in WT generators, representing 59% of the total share, followed by stator faults with 28% and then rotor faults with 9% (the remaining 4% assigned to 'others'). These faults produce excessive heating, flux unbalance, voltages and/or phase current unbalance, decreased average torque, etc., thus negatively affecting the generator's efficiency [57]. Fault frequency components related to such faults are well known and have been demonstrated in different industries in different kinds of induction machines, including a few specific cases in WTs very recently [58–60]. The formulae to obtain the above-mentioned fault frequency components are presented in the following sub-sections.

2.1 Broken rotor bars, f_{BRB}

Broken rotor bars produce an asymmetry in the rotor electric and magnetic fields. As a result, frequency components appear in the stator current that generate speed fluctuation interacting with the supply frequency [61]. Fault frequency components in the presence of broken rotor bars can be calculated as [61–63]

$$f_{BRB} = f_s \pm 2sf_s \tag{1}$$

with f_s the supply frequency and s the slip.

It must be noted that certain degree of asymmetry will always be present in any induction machine (e.g. introduced in manufacturing); thus, the only presence of the components described in Eq. (1) may not indicate such fault. In order to differentiate between fault and no fault, the authors in [63–65] have identified the minimum difference between the amplitudes of such frequency components and the supply frequency that must be found.

Frequency components given by Eq. (1) appear near the supply frequency and overlap with different types of faults. For these reasons, different frequency components have been developed by different authors that facilitate the identification of broken rotor bars, given by [66–68]

$$f_{BRB} = f_s \left[\kappa \left(\frac{1-s}{p} \right) \pm s \right] \tag{2}$$

with κ the harmonic index ($\kappa = 1, 2, 3, \ldots$) and p the number of pole pairs. Based on typical winding configurations, the authors in [52, 69] stated the relation $\kappa/p = 1, 5, 7, \ldots$.

2.2 Stator winding, f_{SW}

Stator winding faults normally develop from inter-turn short circuits [68, 70], where a negative magnetomotive force appears in the windings. As a result, the air-gap flux changes, thus inducing harmonic frequencies in the stator current. Fault frequency components related to stator winding fault are given by [68, 70, 71]

$$f_{SW} = f_s \left[\kappa \left(\frac{1-s}{p} \right) \pm n \right] \tag{3}$$

with $n = 1, 3, 5, \ldots$.

2.3 Bearing faults, f_{BE}

The mechanical displacement that takes place in the case of radial movement between rotor and stator originated with defective bearings results in frequency components in the stator current given by [50, 72]

$$f_{BE} = |f_s \pm \kappa f_{o,i}| \tag{4}$$

with $f_{o,i}$ referring to bearing outer and inner race fault, respectively, calculated as [52, 73, 74]

$$f_o = 0.4 N_b \frac{f_s(1-s)}{p} \tag{5}$$

$$f_i = 0.6 N_b \frac{f_s(1-s)}{p} \tag{6}$$

with N_b the number of bearing balls. These equations are simplifications valid for 8–12 ball bearings.

2.4 Gearbox faults, f_{GBX}

Gearbox damage can be of different nature, originating from a gearbox bearing, shaft, gear or pinion or a combination of them [53]. Therefore, different frequency components associated with a faulty gearbox can appear in the current spectrum. The authors in [75] identified frequency components related to damaged teeth, scoring and debris as

$$f_{GBX} = f_s \left(1 \pm \frac{\kappa}{G_r p}\right) \tag{7}$$

with G_r the gearbox ratio.

The most typical gearbox faults, however, are those related to shaft and gear faults [53, 76–80], calculated as Eqs. (8) and (9), respectively:

$$f^i_{FShaft} = f_s \pm \kappa f^i_{Shaft} \tag{8}$$

$$f^j_{FMesh} = f_s \pm \kappa f^j_{Mesh} \tag{9}$$

with f^i_{Shaft} the gearbox shaft rotating frequencies (with $i = 1, 2, 3 \ldots$) and f^i_{Mesh} the gear mesh frequency (with $j = 1, 2, 3, \ldots, i-1$).

2.5 Eccentricity, f_{ECC}

There are two main groups of fault frequency components associated to eccentricity faults [81]: high- and low-frequency components (HF and LF, respectively). HF components are calculated as [69]

$$f_{ECC,HF} = f_s \left[\pm\nu + (\kappa R \pm n_d)\frac{1-s}{p}\right] \tag{10}$$

with R the number of rotor slots, n_d the eccentricity order ($n_d = 0$ for static eccentricity and $n_d = 1, 2, 3, \ldots$ for dynamic eccentricity) and ν the order of stator time harmonics ($\nu = 1, 3, 5, 7, \ldots$).

Load torque oscillations and load variations do not affect these HF components as much as they affect LF components [81], and they can also separate the spectral components of air-gap eccentricity from broken rotor bars. However, it is necessary to know specific information of the induction machine under study [69].

LF components, on the other hand, appear near the supply frequency, and its magnitude increases in the case of mixed eccentricity. They can be calculated as [69]

$$f_{ECC,LF} = f_s \pm \kappa f_s \frac{1-s}{p} \tag{11}$$

2.6 Rotor asymmetry, f_{RFS}

Rotor mechanical asymmetries such as broken rotor bars, faulty end rings, etc. give rise to new frequency components in the stator current spectra [82–84] that can be calculated as

$$f_{RFS} = f_s(1 \pm 2\kappa s) \tag{12}$$

2.7 Rotor unbalance, f_{RU}

The authors in [85, 86] present a comprehensive analysis of the origins of rotor winding or brush gear unbalance faults, developing their associated analytical expressions. In their study, they identify separately health- and fault-related frequency components, calculated as Eqs. (13) and (14), respectively:

$$f_{HRU} = f_s|6\kappa(1-s) \pm l| \quad (13)$$

$$f_{FRU} = f_s\left|\frac{\kappa}{p}(1-s) \pm l\right| \quad (14)$$

with κ the harmonic constant for air-gap field space ($\kappa = 1, 2, 3, \ldots$) and l the supply time harmonic one ($l = \pm 1$).

2.8 Summary of fault-related frequencies and discussion

A summary of all fault-related frequency components is provided in **Table 2**.

The fault frequency components associated to the most common faults and their possible root causes have been presented. As can be seen, certain defects might produce other defects with the same effect on the current spectra. That is the case for the change in the air-gap flux caused by a fault in the stator winding (given by Eq. (3)), resulting in air-gap eccentricity (given by Eq. (11)), i.e. substituting Eq. (3) with $n = 1$ gives Eq. (11).

In the same way, stator winding impedances produce asymmetries that cause a resultant backward rotating field, thus affecting the rotor currents [73]; this is to say that faults originating in the stator will influence the rotor and vice versa. In fact, deriving from Eq. (3), the same components are obtained than those given by Eq. (14) with $l = 1$.

Further to fault-related frequency components, a number of frequency components that are non-fault related will appear in the stator current spectra of a healthy machine. That would be the case for the stator carrier frequency (i.e. the frequency

Fault	Formula	Eq. label	Constants
Broken rotor bar	$f_{BRB} = f_s \pm 2sf_s$	(1)	$\frac{\kappa}{p} = 1, 5, 7$
	$f_{BRB} = f_s\left[\kappa\left(\frac{1-s}{p}\right) \pm s\right]$	(2)	
Stator winding	$f_{SW} = f_s\left[\kappa\left(\frac{1-s}{p}\right) \pm n\right]$	(3)	$\kappa = 1, 2, 3$
			$n = 1, 3, 5$
Bearing damage	$f_{BE} = \lvert f_s \pm \kappa f_{o,i}\rvert$	(4)	$\kappa = 1, 2, 3$
	$f_o = 0.4N_b\frac{f_s(1-s)}{p}$	(5)	
	$f_i = 0.6N_b\frac{f_s(1-s)}{p}$	(6)	
Gearbox damage	$f_{GBX} = f_s \pm \kappa\left[\frac{f_s}{G_r p}\right]$	(7)	$\kappa = 1, 2, 3$
	$f^i_{FShaft} = f_s \pm \kappa f^i_{Shaft}$	(8)	
	$f^j_{FMesh} = f_s \pm \kappa f^j_{Mesh}$	(9)	
Air-gap eccentricity	$f_{ECC,HF} = f_s\left[\pm\nu + (\kappa R \pm n_d)\frac{1-s}{p}\right]$	(10)	$n_d = 0, 1, 2, 3$
			$\nu = 1, 3, 5, 7$
	$f_{ECC,LF} = f_s\left(1 \pm \kappa\frac{1-s}{p}\right)$	(11)	$\kappa = 1, 2, 3$
Rotor asymmetry	$f_{RFS} = f_s(1 \pm 2\kappa s)$	(12)	$\kappa = 1, 2, 3$
Rotor unbalance	$f_{HRU} = f_s\lvert 6\kappa(1-s) \pm l\rvert$	(13)	$\kappa = 1, 2, 3$
	$f_{FRU} = f_s\left\lvert\frac{\kappa}{p}(1-s) \pm l\right\rvert$	(14)	$l = 1, 2, 3$

Table 2.
CSA formulae.

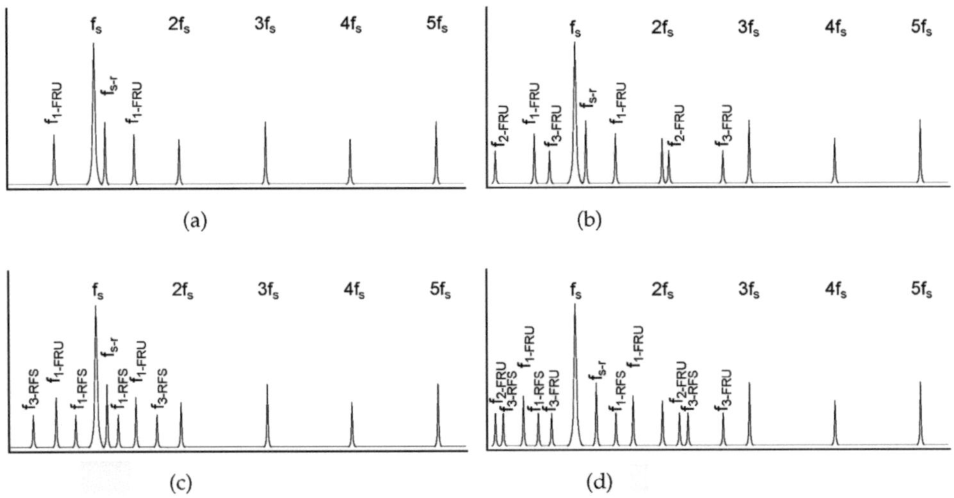

Figure 5.
Spectrum patterns of DFIG under healthy conditions and different fault conditions. (a) Healthy, (b) Rotor Electrical Unbalance, (c) Rotor Mechanical Unbalance, and (d) Rotor Winding Fault.

of the grid, f_s) and its odd and even harmonics ($2f_s$, $3f_s$, $4f_s$, $5f_s$ ...), which are normally present in the spectrum but are not indicative of fault. Another component that will be present in the case of a DFIG, and is not a fault indicator, is that corresponding to the difference between the stator carrier frequency (f_s) and the rotor carrier frequency (f_{rotor}) [58, 59]. This component will appear to the right of f_s when the DFIG operates at a super-synchronous speed (with negative slip). Contrarily, it will appear to the left of f_s at sub-synchronous speed (positive slip). Finally, as previously mentioned, frequency components given by Eq. (13) are not indicators of fault either.

All this highlights the fact that stator current spectra interpretation is not straightforward. Furthermore, fault frequency components are dependent on the rotational speed; thus the potential fault frequency components must be calculated for each operating condition. In order to exemplify in a schematic way typical CSA patterns, some are depicted in **Figure 5**, including that of a healthy DFIG spectra.

3. Case studies

This section presents the diagnosis of three different in-service WTs equipped with DFIGs using CSA. The WTs were located in different sites, thus belonging to different wind farms. Their rated power, number of pole pairs and generator technology vary from one another, as well as the monitoring period and sampling parameters. **Table 3** summarises the rated power (in kW), the monitoring period

	DFIG1	DFIG2	DFIG3
Nominal power [kW]	850	1500	1500
Monitoring period [months]	8	17	1
Sampling frequency [kHz]	1.5	1.5	12
Sampling time [s]	5.4	5.4	10.9

Table 3.
Main characteristics of the three databases used for the analysis.

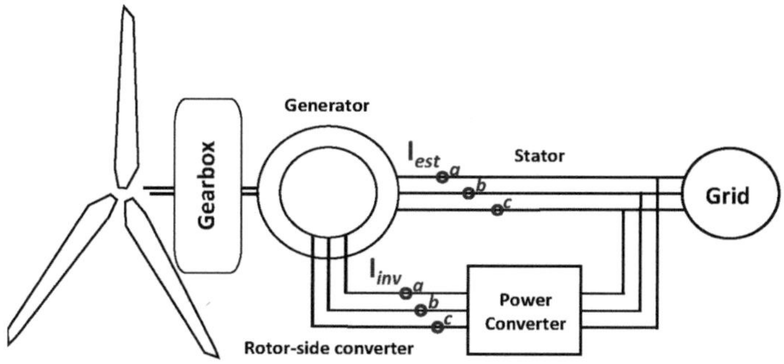

Figure 6.
Drive train diagram with location of sensors in the DFIG.

Sensor location	Label
Stator current phase *a*	I_{est}
Stator current phase *b*	
Stator current phase *c*	
Rotor-side converter current phase *a*	I_{inv}
Rotor-side converter current phase *b*	
Rotor-side converter current phase *c*	

Table 4.
Type of signals used for the analysis.

(in months) and the sampling parameters (sampling rate, in Hertz, and sampling time, in seconds) for each WT DFIG presented in this study.

The data used for the various analyses were extracted from three databases. These comprise currents from sensors located in the induction generator, the stator side and the rotor-side converter, as depicted in **Figure 6**. The electrical measurements used according to **Figure 6** are summarised in **Table 4**.

3.1 Analysis of DFIG1

DFIG1 has two pole pairs, with 1500 rpm rated speed. A test case with the WT operating nearly at full load (825 kW at 1635 rpm) with −0.0915 pu slip was chosen to illustrate the analysis. The negative slip sign is the convention for super-synchronous speed. Under these operating conditions, all possible fault frequencies (as per **Table 2**) were calculated, and peak search and identification were performed in the current spectra, as shown in **Figure 7**.

The supply frequency (50 Hz) and its harmonics (both odd and even) are clearly seen in the spectrum. Another peak naturally present in the spectrum is that corresponding to the difference of the main frequency of the stator and the main frequency of the rotor (f_{s-r} at 54.56 Hz). Then, up to four pairs of f_{FRU} can be seen, as well as the first pair of f_{HRU}. Thus, based on the peaks observed in the spectrum, the presence of rotor electrical unbalance can be concluded.

3.2 Analysis of DFIG2

DFIG2 has three pole pairs, with 1000 rpm rated speed. It presented anomalous vibration levels in the drive-end generator bearing. With only this information, the

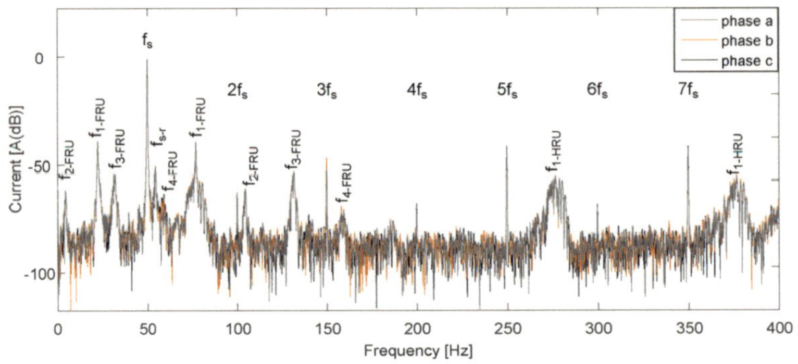

Figure 7.
Three-phase current spectra of DFIG1 showing rotor electrical unbalance.

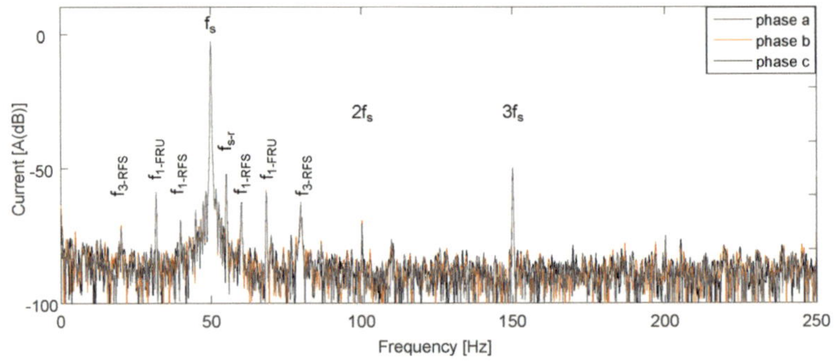

Figure 8.
Three-phase current spectra of DFIG2 showing rotor mechanical unbalance.

wind farm operator decided to replace the generator bearing. Right after the replacement, the vibration levels decreased. However, a few days later, these levels rose to values similar to those prior to the replacement. Visual inspection of the apparently defective bearing showed no fault in the replaced component, thus, the generator bearing was misdiagnosed.

A measurement with the WT operating at super-synchronous speed (1100 rpm, at 1100 kW power) with −0.01 pu slip was selected for this test case. Once again, the set of potential fault frequencies is calculated, and peak search and identification are carried out on the current spectra, as shown in **Figure 8**.

The supply frequency (f_s) corresponds to the highest peak amplitude (at 50 Hz), as expected. It is odd harmonics are also found (150, 250 Hz, etc.), as well as the difference of the stator and rotor carrier frequencies (f_{s-r}). With regard to fault-related frequency components, no peaks related to bearing fault were found (nor to gearbox either). Only the first pair of f_{FRU} can be identified (thus not fault indicator). Finally, two odd pairs of f_{RFS} are seen, belonging to rotor mechanical asymmetries. The diagnosis achieved was, therefore, rotor unbalance of mechanical nature, i.e. unbalance of the high-speed shaft.

3.3 Analysis of DFIG3

DFIG3 has two pole pairs and 1500 rpm rated speed. It was reported with excessive temperature on the rotor windings. Besides, the generator bearings were replaced two or three times a year. With a 3- to 5-year expected lifetime for these

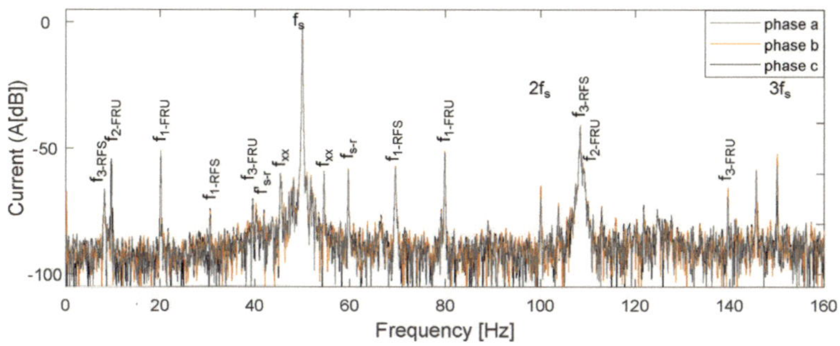

Figure 9.
Three-phase current spectra of DFIG3 showing rotor winding fault.

particular types of generator bearings, the idea that they were suffering damage from a generator fault arose. Thus, CSA was undertaken to find out the health status of the generator.

Similarly to the previous studies, the whole set of potential fault frequency components, as per the formulae previously described, was calculated, and peak search and identification were carried out. To illustrate the analysis, one stationary measurement with the WT operating at 1245 kW and 1791 rpm with per unit slip -0.1941 was selected. Its spectral analysis is shown in **Figure 9**.

Like in the previous analyses, odd and even supply frequency harmonics appear, showing, in this case, certain differences per phase for the second and third harmonics ($2f_s$ and $3f_s$, respectively). Same occurs with the frequency component associated to the difference between the stator carrier frequency (f_s) and the rotor carrier frequency (f_{rotor}), indicated as f_{s-r} in **Figure 9**. Since the measurement was selected with the WT operating at super-synchronous speed, it falls to the right of f_s, as expected. In this case, however, its reflection (denoted as f'_{s-r}) can also be seen to the left of f_s. A minimal amplitude for f'_{s-r} can sometimes be seen; however, the amplitude observed in **Figure 9** for f'_{s-r} on phases a and b is higher than expected. One pair noted as f_{xx} was also found, which does not belong to any of the formulae previously presented.

Regarding fault-related frequency components, three pairs were identified for f_{FRU}. Since more than one pair has been found, it does indicate a mixed eccentricity fault. Finally, two odd pairs of f_{RFS} (f_{1-RFS} and f_{3-RFS}) appear in the current spectra, indicating a fault originating in the rotor. Unlike f_{FRU}, no f_{RFS} pairs are expected to appear in a healthy machine. Hence, it is possible to conclude that DFIG3 has a rotor fault which has led to mixed eccentricity.

4. Conclusions

With the rapid growth of wind energy and development of WTs, researchers and the industry are facing continuous challenges. There is also the fact that the accessibility of offshore wind farms can be limited or restricted for several months. All this, together with the fact that an important share of the existing wind turbine fleet has already achieved its 20-year estimated lifetime, shows that availability and reliability of wind turbines must be developed in parallel. Furthermore, wind energy is expected to play a key role in the short- and long-term electricity market. Thus, the various technical and economical issues surrounding WTs and wind farms must be studied and addressed towards financially viable wind energy.

O&M of wind farms represents an important share of the total expenditure costs; therefore, it is vital to optimise the strategies implemented while increasing reliability and availability of wind turbines. The induction generator is one of the critical components of a WT, together with the drive train and hub and blade assemblies. Under this scenario, the use of electrical measurements to monitor the induction generator has been presented in this chapter.

CSA has been chosen as the CM method for diagnosing three DFIGs of in-service WTs. The technique has been presented, and spectral analyses have been conducted. Fault frequency components related to different faults have been identified in the spectra of the three machines. DFIG1 was diagnosed with rotor electrical unbalance, DFIG2 with rotor mechanical unbalance and DFIG3 with rotor winding fault. CSA has thus been proven as a valid method to diagnose in-service WTs equipped with DFIG that can be implemented in commercial CM systems of WTs.

Acknowledgements

This research originated from the AWESOME Project, funded by the European Union Horizon 2020 research and innovation program under the Marie Sklodowska-Curie Grant Agreement No. 642108.

Nomenclature

CBM	condition-based maintenance
CM	condition monitoring
CSA	current signature analysis
DFIG	doubly-fed induction generator
FBM	failure-based maintenance
GWEC	global wind energy council
LCOE	levelized cost of electricity
LTE	lifetime extension
O&M	operation and maintenance
RTU	remaining useful life
SCADA	supervisory control and data acquisition
TBM	time-based maintenance
WT	wind turbine

Author details

Estefania Artigao*, Andrés Honrubia-Escribano, Sergio Martín-Martínez and Emilio Gómez-Lázaro
Renewable Energy Research Institute and DIEEAC-ETSII-AB, Universidad de Castilla-La Mancha, Albacete, Spain

*Address all correspondence to: estefania.artigao@uclm.es

References

[1] Ohlenforst K. Global wind report 2018. Tech. Rep. Global Wind Energy Council; 2019

[2] Honrubia-Escribano A, Gomez-Lazaro E, Fortmann J, Sorensen P, Martin-Martinez S. Generic dynamic wind turbine models for power system stability analysis: A comprehensive review. Renewable and Sustainable Energy Reviews. 2018;**81**(2):1939-1952

[3] Ren G, Liu J, Wan J, Guo Y, Yu D. Overview of wind power intermittency: Impacts, measurements, and mitigation solutions. Applied Energy. 2017;**204**: 47-65

[4] Wang R-J, Gerber S. Magnetically geared wind generator technologies: Opportunities and challenges. Applied Energy. 2014;**136**:817-826

[5] Burt M, Firestone J, Madsen JA, Veron DE, Bowers R. Tall towers, long blades and manifest destiny: The migration of land-based wind from the Great Plains to the thirteen colonies. Applied Energy. 2017;**206**:487-497

[6] Gao C, Sun M, Geng Y, Wu R, Chen W. A bibliometric analysis based review on wind power price. Applied Energy. 2016;**182**:602-612

[7] Ribrant J, Bertling L. Survey of failures in wind power systems with focus on Swedish wind power plants during 1997–2005. In: IEEE Power Engineering Society General Meeting; 2007. pp. 1-8

[8] Spinato F, Tavner PJ, Bussel GJW, Koutoulakos E. Reliability of wind turbine subassemblies. IET Renewable Power Generation. 2008;**3**(4):387-401

[9] Hahn B, Durstewitz M, Rohrig K. Reliability of wind turbines. Journal of Wind Energy. 2007:329-332

[10] de Prada Gil M, Gomis-Bellmunt O, Sumper A. Technical and economic assessment of offshore wind power plants based on variable frequency operation of clusters with a single power converter. Applied Energy. 2014;**125**: 218-229

[11] Koutoulakos E. Wind turbine reliability characteristics and offshore availability assessment [Master's thesis]. Delft, The Netherlands: Delft University, Wind Energy Research Institute; 2010

[12] Reder MD, Gonzalez E, Melero J. Wind turbine failures-tackling current problems in failure data analysis. Journal of Physics: Conference Series. 2016;**753**:072027

[13] Lin Y, Tu L, Liu H, Li W. Fault analysis of wind turbines in China. Renewable and Sustainable Energy Reviews. 2016;**55**:482-490

[14] Carroll J, McDonald A, McMillan D. Failure rate, repair time and unscheduled O&M cost analysis of offshore wind turbines. Wind Energy. 2015;**19**:1107-1119

[15] Wilson G, McMillan D. Assessing wind farm reliability using weather dependent failure rates. Journal of Physics: Conference Series. 2014;**524**(1): 012181

[16] Hines VA, Ogilvie AB, Bond CR. Continuous reliability enhancement for wind (CREW) database: Wind plant reliability benchmark. Tech. Rep. Sandia National Laboratories, Albuquerque, New Mexico/Livermore, California; 2013

[17] Dinwoodie FQ, McMillan D. Analysis of offshore wind turbine operation and maintenance using a novel time domain meteo-ocean

modelling approach. In: ASME Turbo Expo 2012: Turbine Technical Conference and Exposition. American Society of Mechanical Engineers; 2012

[18] Wilkinson M, Harman K, Hendriks B, Spinato F, van Delft T, Garrad G, et al. Measuring wind turbine reliability—Results of the Reliawind project. In: European Wind Energy Association Conference; 2011. pp. 1-8

[19] Stenberg A, Holttinen H. Analysing failure statistics of wind turbines in Finland. In: European Wind Energy Conference; 2010

[20] Tavner PJ, Xiang J, Spinato F. Reliability analysis for wind turbines. Wind Energy. 2007;**10**:1-18

[21] Sinha Y, Steel J. A progressive study into offshore wind farm maintenance optimisation using risk based failure analysis. Renewable and Sustainable Energy Reviews. 2015;**42**:735-742

[22] Blanco MI. The economics of wind energy. Renewable and Sustainable Energy Reviews. 2009;**13**(6):1372-1382

[23] Fischer K, Besnard F, Bertling L. Reliability-centered maintenance for wind turbines based on statistical analysis and practical experience. IEEE Transactions on Energy Conversion. 2012;**27**(1):184-195

[24] Márquez FPG, Tobias AM, Pérez JMP, Papaelias M. Condition monitoring of wind turbines: Techniques and methods. Renewable Energy. 2012;**46**:169-178

[25] Qiao W, Zhang P, Chow M-Y. Condition monitoring, diagnosis, prognosis, and health management for wind energy conversion systems. IEEE Transactions on Industrial Electronics. 2015;**62**(10):6533-6535

[26] Tian Z, Jin T, Wu B, Ding F. Condition based maintenance optimization for wind power generation systems under continuous monitoring. Renewable Energy. 2011;**36**(5): 1502-1509

[27] Bussel GJW, Zaaijer MB. Reliability, availability and maintenance aspects of large-scale offshore wind farms, a concepts study. In: MAREC 2011: Proceedings of the 2-Day International Conference on Marine Renewable Energies (Institute of Marine Engineers Edition); 2001

[28] Lu B, Li Y, Wu X, Yang Z. A review of recent advances in wind turbine condition monitoring and fault diagnosis. In: IEEE Power Electronics and Machines in Wind Applications (PEMWA); 2009. pp. 1-7

[29] Sheng S. Improving component reliability through performance and condition monitoring data analysis. Tech. Rep. Wind Farm Data Management & Analysis North America; 2015

[30] McMillan D, Ault GW. Towards quantification of condition monitoring benefit for wind turbine generators. In: European Wind Energy Conference Exhibition; Milan, Italy; 2007

[31] Jardine AKS, Lin D, Banjevic D. A review on machinery diagnostics and prognostics implementing condition-based maintenance. Mechanical Systems and Signal Processing. 2006;**20**: 1483-1510

[32] Yam R, Tse P, Li L, Tu P. Intelligent predictive decision support system for condition-based maintenance. The International Journal of Advanced Manufacturing Technology. 2001;**17**(5): 383-391

[33] Nilsson J, Bertling L. Maintenance management of wind power systems

using condition monitoring systems—Life cycle cost analysis for two case studies. IEEE Transactions on Energy Conversion. 2007;**22**(1):223-229

[34] Qiao W, Lu D. A survey on wind turbine condition monitoring and fault diagnosis—Part II: Signals and signal processing methods. IEEE Transactions on Industrial Electronics. 2015;**62**(10): 6546-6557

[35] Hameed Z, Hong Y, Cho Y, Ahn S, Song C. Condition monitoring and fault detection of wind turbines and related algorithms: A review. Renewable and Sustainable Energy Reviews. 2009; **13**(1):1-39

[36] Qiao W, Lu D. A survey on wind turbine condition monitoring and fault diagnosis—Part I: Components and subsystems. IEEE Transactions on Industrial Electronics. 2015;**62**(10): 6536-6545

[37] Kabir MJ, Oo AM, Rabbani N. A brief review on offshore wind turbine fault detection and recent development in condition monitoring based maintenance system. In: IEEE Power Engineering Conference AUPEC; 2015. pp. 1-7

[38] Coronado D, Fischer K. Condition monitoring of wind turbines: State of the art, user experience and recommendations. Tech. Rep. Fraunhofer Institute for Wind Energy and Energy System Technology IWES Northwest; 2015

[39] Yang W, Tavner PJ, Crabtree C, Feng Y, Qiu Y. Wind turbine condition monitoring: Technical and commercial challenges. Wind Energy. 2012;**17**(5): 673-693

[40] Maldonado J, Alvarez O. A review of concepts and methods for wind turbines condition monitoring. In: IEEE Computer and Information Technology (WCCIT), 2013 World Congress; 2014

[41] Amirat Y, Benbouzid MEH, Al-Ahmar E, Bensaker B, Turri S. A brief status on condition monitoring and fault diagnosis in wind energy conversion systems. Renewable and Sustainable Energy Reviews. 2009;**13**(9):2629-2636

[42] Popa LM, Jensen B-B, Ritchie E, Boldea I. Condition monitoring of wind generators. In: IEEE Industry Applications Conference, 38th IAS Annual Meeting; 2003. pp. 1839-1846

[43] Singh GK, Kazzaz S. Induction machine drive condition monitoring and diagnostic research—A survey. Electric Power Systems Research. 2003;**64**(2): 145-158

[44] Tautz-Weinert J, Watson SJ. Using SCADA data for wind turbine condition monitoring—A review. IET Renewable Power Generation. 2016;**11**(4):382-394

[45] Gonzalez E, Tautz-Weinert J, Melero J, Watson SJ. Statistical evaluation of SCADA data for wind turbine condition monitoring and farm assessment. Journal of Physics: Conference Series. 2018;**1037**:032038

[46] Warner E. Global Wind Report. Annual market update. Tech. Rep. Global Wind Energy Council; 2015

[47] Rubert T, McMillan D, Niewczas P. A decision support tool to assist with lifetime extension of wind turbines. Renewable Energy. 2017;**120**(1):423-433

[48] Ziegler L, Gonzalez E, Rubert T, Smolka U, Melero JJ. Lifetime extension of onshore wind turbines: A review covering Germany, Spain, Denmark, and the UK. Renewable and Sustainable Energy Reviews. 2017;**82**:1261-1271

[49] Tavner PJ. Review of condition monitoring of rotating electrical machines. IET Electric Power Applications. 2007;**2**(4):215-247

[50] Benbouzid MEH. A review of induction motors signature analysis as a medium for faults detection. IEEE Transactions on Industrial Electronics. 2000;**47**(5):984-993

[51] Siddique A, Yadava G, Singh B. A review of stator fault monitoring techniques of induction motors. IEEE Transactions on Energy Conversion. 2005;**20**(1):106-114

[52] Benbouzid MEH, Kliman GB. What stator current processing-based technique to use for induction motor rotor faults diagnosis? IEEE Transactions on Energy Conversion. 2003;**18**(2):238-244

[53] Mohanty A, Kar C. Fault detection in a multistage gearbox by demodulation of motor current waveform. IEEE Transactions on Industrial Electronics. 2006;**53**(4): 1285-1297

[54] Blodt M, Granjon P, Raison B, Rostaing G. Models for bearing damage detection in induction motors using stator current monitoring. IEEE Transactions on Industrial Electronics. 2008;**55**(4):1813-1822

[55] Shahriar MR, Borghesani P, Tan AC. Electrical signature analysis-based detection of external bearing faults in electromechanical drivetrains. IEEE Transactions on Industrial Electronics. 2018;**65**(7):5941-5950

[56] Alewine K, Chen W. A review of electrical winding failures in wind turbine generators. IEEE Electrical Insulation Magazine. 2012; **28**(4):8-13

[57] Yazidi A, Henao H, Capolino G, Artioli M, Filippetti F, Casadei D. Flux signature analysis: An alternative method for the fault diagnosis of induction machines. In: IEEE Powertech; 2005. pp. 1-6

[58] Artigao E, Honrubia-Escribano A, Gomez-Lazaro E. Current signature analysis to monitor DFIG wind turbine generators: A case study. Renewable Energy. 2018;**116**:5-14

[59] Artigao E, Koukoura S, Honrubia-Escribano A, Carroll J, McDonald A, Gómez-Lázaro E. Current signature and vibration analyses to diagnose an in-service wind turbine drive train. Energies. 2018;**11**(4):960

[60] Cheng F, Qu L, Qiao W, Wei C, Hao L. Fault diagnosis of wind turbine gearboxes based on DFIG stator current envelope analysis. IEEE Transactions on Sustainable Energy. 2018;**10**:1044

[61] Siau J, Graff A, Soong W, Ertugrul N. Broken bar detection in induction motors using current and flux spectral analysis. Australian Journal of Electrical and Electronics Engineering. 2004;**1**(3):171-178

[62] Acosta G, Verucchi C, Gelso E. A current monitoring system for diagnosing electrical failures in induction motors. Mechanical Systems and Signal Processing. 2006;**20**(4): 953-965

[63] Siddiqui KM, Giri VK. Broken rotor bar fault detection in induction motors using transient current analysis. International Journal of Electronics and Communication Technology. 2011;**2**(4): 114-119

[64] Nandi S, Toliyat HA, Li X. Condition monitoring and fault diagnosis of electrical motors—A review. IEEE Transactions on Energy Conversion. 2005;**20**(4):719-729

[65] Bellini A, Filippetti F, Franceschini G, Tassoni C. Quantitative evaluation of induction motor broken bars by means of electrical signature analysis. IEEE Transactions on Industry Applications. 2001;**37**(5):1248-1255

[66] Kliman GB, Koegl RA, Stein J, Endicott RD, Madden MW. Noninvasive detection of broken rotor bars in operating induction motors. IEEE Transactions on Energy Conversion. 1988;**3**(4):873-879

[67] Rodriguez PVJ, Negrea M, Arkkio A. A simplified scheme for induction motor condition monitoring. Mechanical Systems and Signal Processing. 2008;**22**(5):1216-1236

[68] JJung JH, Lee JJ, Kwon BH. Online diagnosis of induction motors using MCSA. IEEE Transactions on Industrial Electronics. 2006;**53**(6):1842-1852

[69] Kliman GB, Stein J. Methods of motor current signature analysis. Electric Machines and Power Systems. 1992;**20**(5):463-474

[70] Joksimovic GM. The detection of inter-turn short circuits in the stator windings of operating motors. IEEE Transactions on Industrial Electronics. 2000;**47**(5):1078-1084

[71] Douglas H, Pillay P, Barendse P. The detection of interturn stator faults in doubly-fed induction generators. In: Fourtieth IAS Annual Meeting. Conference Record of the 2005 Industry Applications Conference; 2005. pp. 1097-1102

[72] Mehala N. Condition monitoring and fault diagnosis of induction motor using motor current signature analysis [PhD thesis]. Kurukshetra, India: National Institute of Technology, Kurukshetra; 2010. Registration No. 2K07-NITK-PhD-1160-E

[73] Thomson WT, Fenger M. Current signature analysis to detect induction motor faults. IEEE Industry Applications Magazine. 2001;**7**(4):26-34

[74] Thomson WT, Gilmore RJ. Motor current signature analysis to detect faults in induction motor drives—Fundamentals, data interpretation and industrial case histories. In: Proceedings of the Thirty-Second Turbomachinery Symposium; 2003

[75] Rajagopalan S, Habetler TG, Harley RG, Sebastian T, Lequesne B. Current/voltage-based detection of faults in gears coupled to electric motors. IEEE Transactions on Industry Applications. 2006;**42**(6):1412-1420

[76] Kia SH, Henao H, Capolino G-A. Gearbox monitoring using induction machine stator current analysis. In: IEEE International Symposium on Diagnostics for Electric Machines, Power Electronics and Drives (SDEMPED). IEEE; 2007. pp. 149-154

[77] Kia SH, Henao H, Capolino G-A. Gear tooth surface damage fault detection using induction machine electrical signature analysis. In: 9th IEEE International Symposium on Diagnostics for Electric Machines, Power Electronics and Drives (SDEMPED). IEEE; 2013. pp. 358-364

[78] Kar C, Mohanty AR. Monitoring gear vibrations through motor current signature analysis and wavelet transform. Mechanical Systems and Signal Processing. 2006;**20**:158-187

[79] Cheng F, Qu L, Qiao W. Fault prognosis and remaining useful life prediction of wind turbine gearboxes using current signal analysis. IEEE Transactions on Sustainable Energy. 2017;**9**:157

[80] Neti P, Zhang P, Shah M, Younsi K. Electrical signature analysis based online monitoring of drive-trains for doubly-fed wind generators. In: 38th Annual Conference on IEEE Industrial Electronics Society (IECON); 2012

[81] Faiz J, Moosavi S. Eccentricity fault detection—From induction machines to DFIG—A review. Renewable and

Sustainable Energy Reviews. 2016;**55**: 169-179

[82] Gritli Y, Stefani A, Rossi C, Filippetti F, Chatti A. Experimental validation of doubly fed induction machine electrical faults diagnosis under time-varying conditions. Electric Power Systems Research. 2011;**81**(3): 751-766

[83] Gritli Y, Zarri L, Rossi C, Filippetti F, Capolino G-A, Casadei D. Advanced diagnosis of electrical faults in wound-rotor induction machines. IEEE Transactions on Industrial Electronics. 2013;**60**(9):4012-4024

[84] Bellini A, Filippetti F, Tassoni C, Capolino G-A. Advances in diagnostic techniques for induction machines. IEEE Transactions on Industrial Electronics. 2008;**55**(12):4109-4126

[85] Djurovic S, Williamson S, Renfrew A. Dynamic model for doubly-fed induction generators with unbalanced excitation, both with and without winding faults. IET Electric Power Applications. 2009;**3**(3):171-177

[86] Williamson S, Djurovic S. Origins of stator current spectra in DFIGs with winding faults and excitation asymmetries. In: IEEE International Electric Machines and Drives Conference, IEMDC'09; Vol. 9; 2009. pp. 563-570

Chapter 9

Reliability Assessment of Wind Turbines

Mai F.M. Ayoub

Abstract

There are a wide variety of wind turbines types. The selection of a wind turbine type, the site of wind turbines fields erection and the maintenance scheme are basic parameters which should be carefully considered for optimum performance and reliable operation and power output. Many techniques had been developed and refined to represent and study the wind turbines complex system in order to make their operation safe, reliable and maintainable. In the present work, emphasis will be given to reliability block diagram quantitative technique to assess the reliability of wind turbines systems. This entails the application of reliability theories on wind turbines individually and wind turbines fields as a whole to ensure maximum utilization of available wind power. Specially devised computer software has been performed and applied on exemplary wind turbine field. The proposed computer program has shown to be helpful if adopted in assessing wind turbine fields giving indications of their reliability.

Keywords: wind turbines, wind farms, reliability, availability, maintenance

1. Introduction

Wind power projects have been developed rapidly in the last two decades due to the increase in fossil-fuel prices. Hence, energy policies have been created for renewable energy research and technical development. Wind power has been developed and became a new fast-growing industry to compete the existing fossil-fueled power plants [1]. Wind turbines have reached nowadays sizes of 8 MW [2] and wind farms are planned at a size of up to 1020 MW. Hence, large arrays of turbines, known as wind farms, become an increasingly important source of renewable energy and are used to reduce dependence on fossil fuels and protecting the environment [3].

The efficiency and the life time of any wind power project depends largely on the operation and maintenance function. This means that it is a primary aim for any manufacturer or investor is to have wind turbines with a very high level of reliability [3]. System reliability is defined as "the probability that the system will perform its intended function for a certain period of time under stated conditions". This requires critical considerations to enable us to decide on the quality and frequency of maintenance required by keeping records of failure [4].

Classical reliability analysis techniques use parameters derived from actual test data in models to evaluate the performance of system or components. The analysis is based on the time-to-failure data of the component, either under use conditions or from accelerated life tests. Hence, a system (a collection of components, subsystems and/or assemblies) is designed in order to achieve desired function with acceptable performance and reliability". Therefore, the type of these components, their quantities, qualities, the manner in which they are arranged in the system and the relationship between these components would affect overall reliability of the system. This reliability relationship is usually expressed by using logical diagrams, such as Reliability Block Diagram (RBD) and/or Fault Trees [5]. The main objective of a reliability study would be to provide information as a basis for decision [6]. So, the results provided by the study of reliability does not tell us exactly what is a decision that we must take, but guide us towards optimum solution. For example, it can be useful in the study of reliability fields, risk analysis, optimization of operations and maintenance. Risk analysis is a way to identify the causes and consequences of failure events. In this case measuring the availability of the device will be more appropriated than reliability, as the availability of a repairable device is defined as: "The proportion of time, during the intended service time, that the device is present or ready for service."

This chapter presents reliability calculation method as software that calculates the reliability of any system with any number of parallel and serial components. This method applied to reliability-based approach to select appropriate wind turbine types for a wind farm considering site-specific wind speed patterns.

2. Reliability analysis techniques

The challenge in complex systems is the problem of analyzing and predicting how reliable they are. These complex systems goal is always to make them safer and more reliable. In order to do this we have to identify the parts that contribute most of the risks involved with their use. Many techniques have been developed and refined in order to more accurately represent these complex systems. Both qualitative and quantitative methods have been developed to analyze complex systems, some are used more widely than others, while some are developed primarily for one specific application, but all techniques have their advantages and disadvantages [7].

2.1 Qualitative techniques

Qualitative reliability analysis methods have been used to identify all possible system failures, and risks of each failure. The most widely used qualitative method is failure modes and effects analysis (FMEA), sometimes also known as failure modes, effects and criticality analysis (FMECA). These are methodologies designed to identify potential failure modes for a product or process, to assess the risk of failures and rank issues in terms of importance and identify the corrective actions. This is generally done by identifying certain characteristics: how each of these parts may fail, what can the result in these failures, what are the possible effects of these failures, how failures can be discovered, and what provisions are provided to compensate for this design failure [7].

FMEA can be completed either on an existing system or during the design phase, and applied at different stages meets different objectives. When done during the design phase, it can help choose design alternatives with high safety and reliability. It can also help develop test planning, which can provide a basis for any quantitative reliability analysis to be performed.

2.2 Quantitative techniques

There are several methods of quantitative reliability analysis techniques, with various theories behind them. The three that are most widely used are fault tree analysis (FTA), reliability block diagrams (RBD), and Markov analysis (MA). Each method of these three methods is the best for different cases. Quantitative analysis depends on the data of how a system works, often gained from previously completed qualitative assessments, and apply information about failure rates, probabilities, characteristics, and so on to this data in order to learn more about subsystems or the system as a whole. Then, based on the method of analysis used, the result is a form of system failure data and can be used to perform a range of tasks, most notably identifying the largest contributors to risk in the system in order to improve them and thereby reduce the risk to the system [7].

These main quantitative reliability techniques can be summarized as follows:

2.2.1 Fault tree

The fault tree shows all possible combinations of failure events that may cause a specific system failure. Fault trees are created by deductively thinking about the cause of failure. Component failures and other events are combined by logical operations "AND" ($\cap$) and "OR" ($\cup$) to provide a logical description of the failure [8].

2.2.2 Reliability block diagram

The reliability block diagram shows how the performance of components or subsystems allows to meet the function of a specific system. These diagrams facilitate the calculation of reliability indicators and illustrate the role of redundancy [9].

2.2.3 Markov analysis

For any given system, the Markov model consists of a list of possible states of the system, possible transition paths between those states, and the parameters of the rate of such transitions. In reliability analysis, transitions usually consist of failures and repairs. When a Markov model is graphically represented, each state is usually depicted as a "bubble," where arrows indicate paths between states [10].

3. Wind turbines reliability and life time

The selection of components to describe the main system is not just an arbitrary choice; it is a choice of what is useful in practice and where available data can be found. The choice of which component should be used to model the entire system should be chosen based on functional and available information (Reliability performance and maintenance – a survey of failures in wind power systems).

As regards wind turbines life time, changes in reliability with increasing operational life can provide indicators of life expectancy and the amount of maintenance required. Reliability can be expressed in the failure rate. The principle of evolution is well-known as bathtub curve: sign of early failures in the beginning of the process followed generally by a longer period of random failures, with a fixed rate statistically, until it starts increasing with age process (Failure - wear out) because of the accumulation of wear and tear [11].

However, the actual total age of a turbine, naturally, differs depending on various technical systems for wind turbines, the loads, the environmental conditions, maintenance plans for each turbine, the operating conditions and labor skills.

4. Reliability computational technique using reliability block diagrams (RBD)

Block diagrams can also be used to describe the relation between components and system definition. When used in this fashion, the block diagram is referred to as the Reliability Block Diagram (RBD). A reliability block diagram is a graphical representation of system components and how they relate to reliability (connected). (Note: One can also think of RBD as a logical scheme of the system based on its characteristics. It should also be noted that this may differ from how the components are actually connected).

After determining the properties of each block in the system, the blocks can then be connected reliably to create a system block reliability diagram. RBD provides a visual representation of the way the blocks are arranged in terms of reliability. This diagram shows the effect of component success or failure on system success or failure.

For example, if all components in a system must succeed in order for the system to succeed, the components will be arranged by reliability in series. If one of the components must succeed for the system to succeed, then these two components are ranked in reliability in parallel.

This order of reliability of the components is directly related to the mathematical description derived from the system. The mathematical description of the system is the key to determining the system reliability. Actually, the reliability of the system is that mathematical description (obtained using probabilistic methods) which determines the reliability of the system in terms of components reliabilities. The result is an analytical expression describing system reliability as a function of time based on the reliability functions of its components.

4.1 Series systems

In a series system **Figure 1**, the reliability of the system is the probability that component 1 succeeds and component 2 succeeds and all of the other components in the system succeed. So all components must succeed for the system to succeed. Then, the system reliability will be given by:

$$R_s = P(X_1 \cap X_2 \cap .....\cap X_n) \tag{1}$$

where:

- Rs is the reliability of the system.
- Xi is the event of component i being operational.
- $P(Xi)$ is probability that component i is operational.

In the case where the failure of a component affects the failure rates of other components (i.e. the life distribution characteristics of the other components

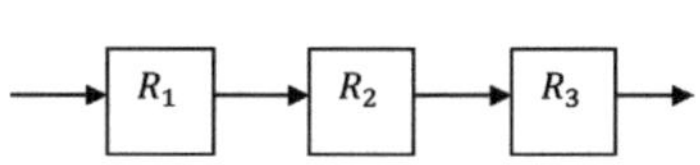

Figure 1.
Simple series systems.

change when one component fails), then the conditional probabilities in equation above must be considered.

However, in the case of independent components, equation above becomes:

$$R_s = R\ (X_1, X_2, X_3, \ldots X_n) = P(X_1)P(X_2)P(X_3)\ldots P(X_n)$$

$$= \prod_{i=1}^{n} P(X_i) \tag{2}$$

Or, in terms of individual component reliability

$$R_s = \prod_{i=1}^{n} R_i \tag{3}$$

In other words, for a pure series system, the system reliability is equal to the product the reliabilities of its constituent components.

4.2 Parallel systems

In a simple parallel system, as shown in the figure below, at least one unit must succeed for the system to succeed. Parallel units are also referred to as redundant units. Redundancy is a very important aspect of system design and reliability as redundancy is one of several ways to improve system reliability.

The probability of failure, or unreliability, for a system that has statistically independent parallel components **Figure 2** is the probability of failure of component 1, failure of component 2 and failure of all other components of the system. So in a parallel system, all n components must fail until the system fails. In other words, if component 1 succeeds, component 2 succeeds, or any component n succeeds, the system will succeed. The system's unreliability (failure rate) is then given by:

$$Q_s = P(X_1 \cap X_2 \cap \ldots\ldots \cap X_n\ ) \tag{4}$$

where:

- $Q\,s$ is the unreliability of the system
- $X\,i$ is the event of failure of unit i
- $P(X\,i)$is probability of failure unit i

In the case where component failure affects failure rates of other components, conditional probabilities must be considered as the above equation. However, in the case of independent components, the above equation becomes:

$$Q_s = P(X_1)P(X_2)P(X_3)\ldots P(X_n) = \prod_{i=1}^{n} P(X_i) \tag{5}$$

Or, in terms of component unreliability:

$$Q_s = \prod_{i=1}^{n} Q_i \tag{6}$$

Observe the variation with the series system, where the reliability of the system was the product of components reliabilities; whereas the parallel system has the overall system unreliability as a product of component unreliability (failure rates).

The reliability of the parallel system is then given by:

$$R_s = 1 - Q_s = 1 - (Q_1 * Q_1 * ... * Q_n) = 1 - [(1 - R_1) * (1 - R_2) * ... * (1 - R_n)]$$

$$R_s = 1 - \prod_{i=1}^{n}(1 - R_i) \quad (7)$$

In a series system, the least reliable component has the biggest effect on the reliability of the system. However, the component with the highest reliability in a parallel system has the biggest effect on the system's reliability, since the most reliable component is the one that will most likely fail last. This is a very important property of the parallel configuration, specifically in the selection, design and improvement of systems.

4.3 Combined series and parallel systems

While many smaller systems can be accurately represented by a simple series or parallel configuration, there may be larger systems that include both parallel and series configurations in the overall system. These systems can be analyzed by calculating the reliability of the individual series and parallel segments and then integrating them in an appropriate manner. This methodology is illustrated in **Figure 3**.

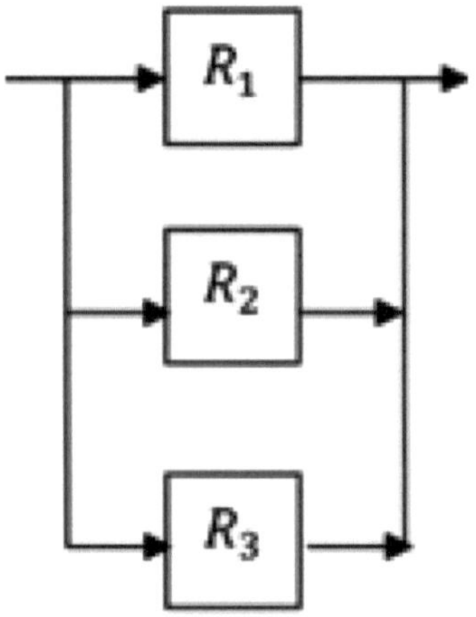

Figure 2.
Simple parallel system.

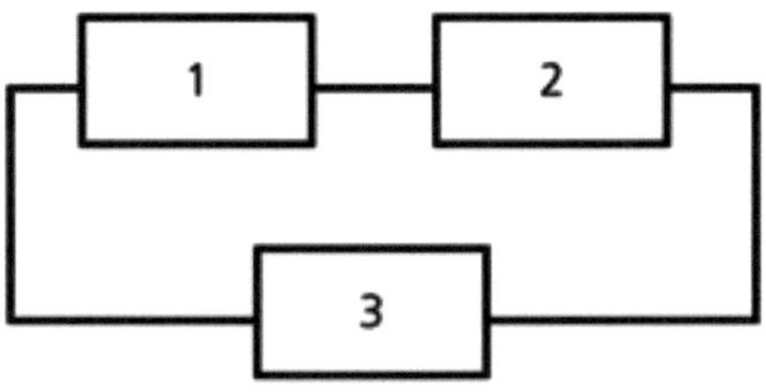

Figure 3.
Combination of series and parallel.

First, the reliability of the series segment consisting of components 1 and 2 is calculated:

$$R_{1,2} = R_1 . R_2 \tag{8}$$

The reliability of the overall system is then calculated by treating component 1 and component 2 as one component with reliability connected in parallel with component 3. Therefore:

$$R_s = 1 - [(1 - R_{1,2}) * (1 - R_3)] \tag{9}$$

5. Reliability assessment and calculation software

Advanced reliability assessment and distribution reliability analysis provides engineers with an efficient and effective tool for estimating the performance of power systems. Using flexible input parameters, results can be quickly obtained for both radial and looped systems. Powerful calculation techniques allow engineers to choose the depth of system design and the associated results.

Reliability software is a technique by which one can draw conclusion about the behavior of the system and its failure rate based on back history or manufacturer given reliability of the components and/or subsystems. There is a lot of reliability

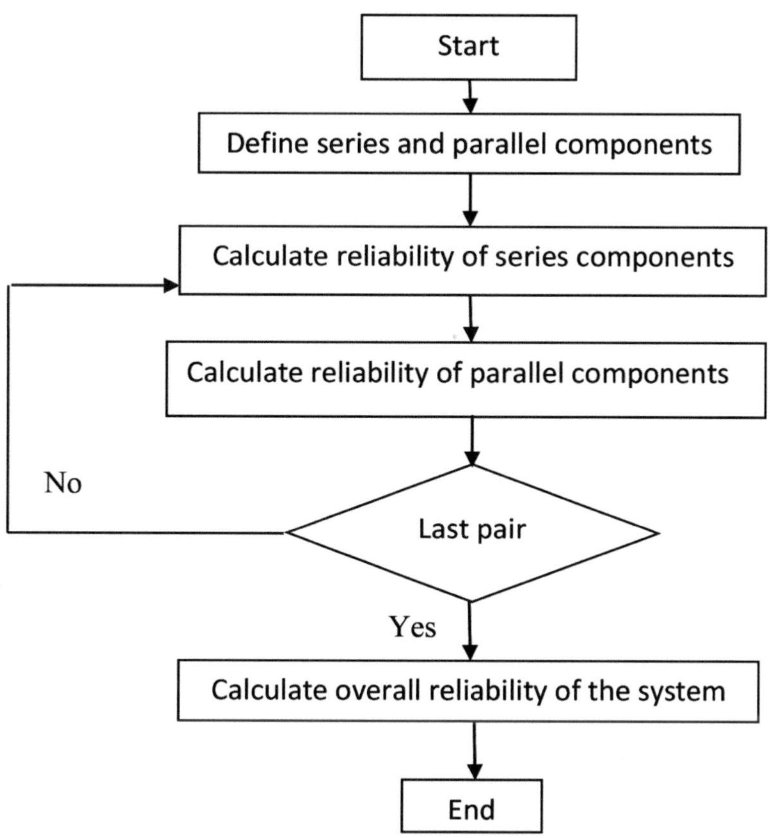

Figure 4.
Reliability calculations program's flow chart.

software in the market; however, the present proposed software represents an endeavor towards devising a computerized simple interactive technique for reliability assessment. The software was programmed using "Visual Basic 6" to develop general reliability calculations software which is suitable for solving any system in simple steps to understand and use.

5.1 Program description

The present program deals with the reliability calculations of any system-with any number of components- connected in parallel and/or in series based on its back history.

Figure 4. present a generalized flow chart of the first version (reliability calculation software), the second version (reliability-based selection of wind turbines) will be discussed later as shown in Section 6.1.

5.2 Program inputs

Program inputs are the reliability block diagram data represented at a text file (.txt) contains the number of components (items) in the top, then the component number, its reliability and its location on the RBD is defined by predecessors and followers as shown in **Figure 5**.

9

Item	R	Predecessors	Followers
1	0.6	0 0 0 0 0 0 0 0 0 0 0	2 3 0 0 0 0 0 0 0 0 0
2	0.6	1 0 0 0 0 0 0 0 0 0 0	4 0 0 0 0 0 0 0 0 0 0
3	0.6	1 0 0 0 0 0 0 0 0 0 0	4 0 0 0 0 0 0 0 0 0 0
4	0.3	2 3 0 0 0 0 0 0 0 0 0	5 0 0 0 0 0 0 0 0 0 0
5	0.6	4 0 0 0 0 0 0 0 0 0 0	6 7 8 0 0 0 0 0 0 0 0
6	0.6	5 0 0 0 0 0 0 0 0 0 0	9 0 0 0 0 0 0 0 0 0 0
7	0.6	5 0 0 0 0 0 0 0 0 0 0	9 0 0 0 0 0 0 0 0 0 0
8	0.6	5 0 0 0 0 0 0 0 0 0 0	9 0 0 0 0 0 0 0 0 0 0
9	0.4	6 7 8 0 0 0 0 0 0 0 0	10 0 0 0 0 0 0 0 0 0 0

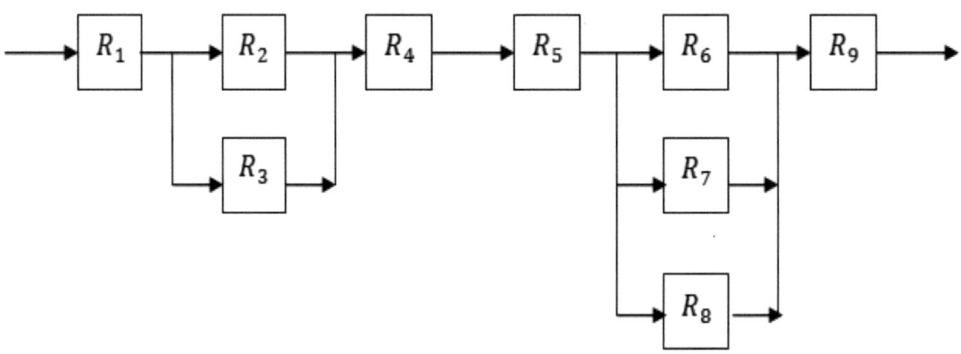

Figure 5.
Program's input reliability block diagram example.

5.3 Program outputs

Program output is the resultant value of the overall reliability of the system.

6. Reliability-based selection of wind turbines software

This application represents reliability/availability-based approach to select appropriate wind turbine types for a wind farm considering site-specific wind speed patterns.

Assuming a constant failure rate, reliability prediction using the exponential distribution, would be most appropriate. If λ is the failure rate and t is the time, then the reliability R(t) can be determined by [12]:

$$R(t) = e^{-\lambda t} \tag{10}$$

The mathematical description of the system is the key to the determination of the reliability of the system. In fact, the system's reliability function is that mathematical description (obtained using probabilistic methods) and it defines the system reliability in terms of the component reliabilities. The result is an analytical expression that describes the reliability of the system as a function of time based on the reliability functions of its components. In the case of the wind turbine all components are in series as shown in **Figure 6** and the overall reliability of the system, R_s, can be determined as follows [13]:

$$R_s = R\ (X_1, X_2, X_3, \ldots X_n) = P(X_1)P(X_2)P(X_3)\ldots P(X_n) \tag{11}$$

where $P(X_n)$ is .. Component reliability.

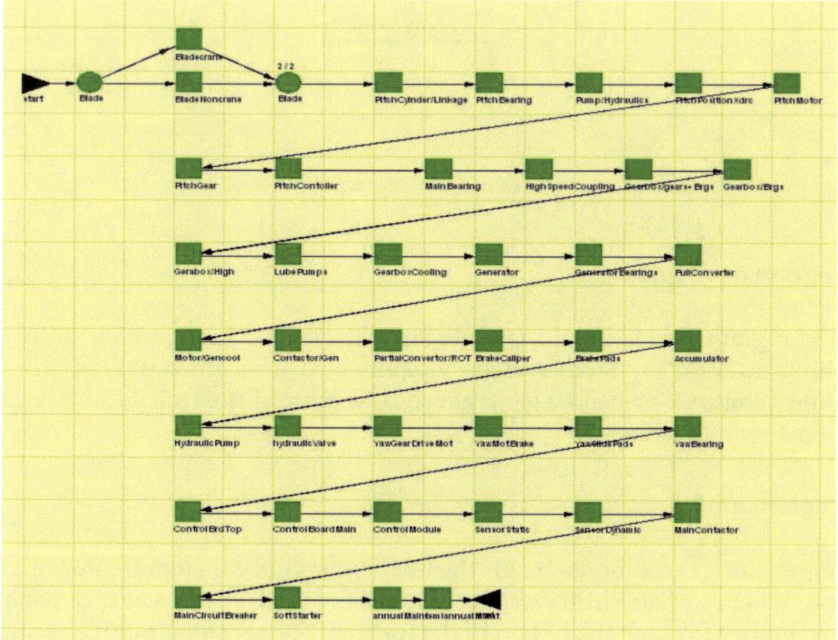

Figure 6.
Wind turbine reliability block diagram [14].

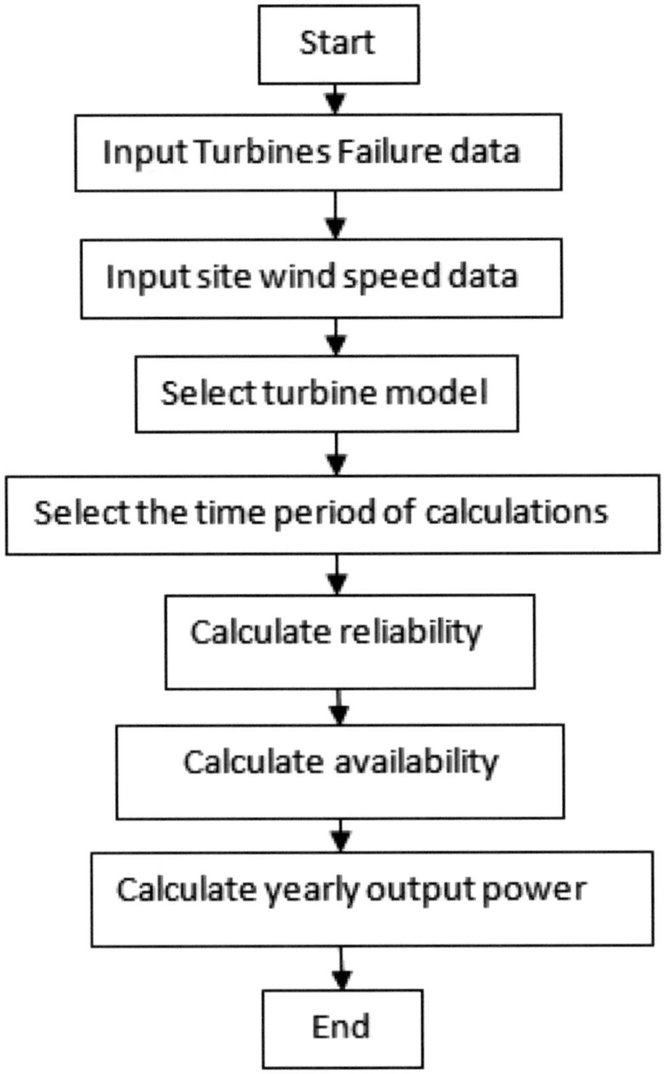

Figure 7.
Reliability/availability based selection of wind turbines. Program's flow chart.

6.1 Program description

This program used to build a module for WT selection based on reliability and availability according to each turbine's components back history and the site's wind availability. **Figure 7** presents a generalized flow chart of the (reliability based selection of wind turbines).

6.2 Program inputs

Program have three inputs first is the turbines failure data represented at a text file (.txt) which contains historical data for both of failure rate/year and mean time to repair in days for each subsystem in the turbine. The second input file contains all turbine's data (number of turbines, turbine's name, Height, Rated Power, wind speeds, Reliability File and Photo File if available). The third input file contains average hourly wind speeds for each site in the study all over the year.

6.3 Program outputs

Program outputs are the followings: site's wind availability all over a year represented as a histogram, wind turbine's reliability and availability for the chosen period of time, and also the yearly output power for the selected turbine in the selected specific site.

Both of version one, "Reliability Calculations software", and version two, "Reliability/Availability Based Selection of Wind Turbines", was tested and compared with the traditional calculation methods. The software has advantage over the analytical approaches since version one can calculate the reliability of any complex system, whatever its components number where or how this components were connected, with a very high speed and perfectly with no need for professional user. This advantage minimized error opportunities and saves lot of work and time. The second version was used to build a module for wind turbines selection based on reliability and availability and was capable of differentiating between different wind turbines systems based on reliability and availability criterion. It was very useful to graph results and understand the effect of reliability of the components with the wind availability of the site on the power output of the turbine. Those graphs also can help the investors to choose the most suitable turbine for the selected site in order to maximize production and minimize operation and maintenance costs. Accordingly the aims of the current study have been satisfactorily fulfilled.

7. Conclusions

The implementation of the reliability/availability calculations software will be very useful for understanding and preparing maintenance schedules. The developed methodology can give the user a proper selection of wind turbines fields according to the back history of wind turbine components and site specifications.

Author details

Mai F.M. Ayoub
Mechanical Engineering Department, National Research Centre, Cairo, Egypt

*Address all correspondence to: maifayez.mf86@gmail.com

References

[1] Ackermann T. Wind Power in Power Systems. 1st ed. New York: John Wiley and Sons; 2005

[2] Kennedy S. Wind power planning: Assessing long term costs and benefits. Energy Policy. 2005;**33**:1661-1675

[3] Wiselius T. Developing Wind Power Projects-Theory & Practice. 1st ed. UK and USA: Earthscane; 2007

[4] Advisory Group on Reliability of Electronic Equipment (AGREE). Reliability of military electronic equipment. In: Task Group 9 Report. Washington, DC: US Government Printing Office; June 1957

[5] Keisner A. Reliability Analysis Technique Comparison as Applied to Space Shuttle. AE 8900 OLD-Special Topic. USA: NASA; 2003

[6] Rausand M, Hoyland A. System Reliability Theory. Hoboken: John Wiley and Sons; 2004. 0-471-47133-X

[7] Wallace R, Blischke DN, Murthy P. Reliability: Modeling, Prediction, and Optimization. New York: Wiley and Sons; 2000

[8] Haasl DF, Roberts NH, Vesely WE, Goldberg FF. Fault Tree Handbook. USA: Nuclear Regularity Commission. NUREG-0492. Available from: http://www.weibull.com/basics/fault-tree/index.html

[9] Applied R and M Manual for Defence Systems (GR-77 Issue 2012), Chapter 30. Available from: http://www.sars.org.uk/; May 2012

[10] Applied R&M Manual for Defence Systems (GR-77 Issue 2012), Chapter 38. Available from: http://www.sars.org.uk/; May 2012

[11] Greco A, Sheng S, Keller J, Erdemir A. Material wear and fatigue in wind turbine Systems. Wear. 2013;**302**(1-2):1583-1591. https://doi.org/10.1016/j.wear.2013.01.060

[12] Portsmouth BS. Introduction to Reliability; April 2012. Available from: woodm.myweb.port.ac.uk/q/reliability.pdf

[13] Lipson C, Sheth NJ. Statistical Design and Analysis of Engineering Experiments. McGraw Hill Education, International Edition; 1973. ISBN: 978-0070854475

[14] Hill RR, Stinebaugh JA, Briand D, Benjamin AS, Linsday J. Wind Turbine Reliability: A Database and Analysis Approach, SAND2008-0983. February: Sandia National Laboratories; 2008

Chapter 10

Wind Energy Conversion System Modeling toward Different Approaches

Ali El Yaakoubi, Lahcen Amhaimar and Adel Asselman

Abstract

The main focus of this chapter is to modeling the different parts of the wind energy conversion system (WECS) and reviewing the different approaches used in this context. The chapter starts with the aerodynamic and the structural modeling of the *wind turbines* (WTs), and a description of the steps used to derive a linear time invariant (LTI) model. Thereafter, the chapter introduces models of the electrical actuators in the three phases (*abc*) and park phases (*dq*) reference frames, and recalls the assumptions considered. The chapter finishes by presenting the *pulse width modulation* (PWM) control strategy, the power converters and the pitch actuator models.

Keywords: wind turbines, aerodynamics modeling, structural modeling, linear time invariant model, wind turbines control

1. Introduction

Wind, which is the main driving source of the wind turbines (WTs), is characterized by speed and direction. Its origin is movements of air masses in the atmosphere due to temperature or pressure differences. In the lower layer of the atmosphere, the winds are delayed by frictional forces and obstacles that change the wind speed and direction. This is the origin of the turbulence flows. Turbulence increases with higher ground roughness and decreases with distance from the ground, which causes the wind speed to vary over a wide range of amplitudes and frequencies. The major characterization of the winds in the lower layer is that the kinetic energy is distributed in the frequency domain, which is known as the Van der Hoven spectrum [1]. According to this model, the kinetic energy is concentrated around two frequencies, which allows splitting the wind speed (v) in two components: a slowly variable deterministic component (v_m), and a rapid variable stochastic component (v_t) as expressed by Eq.(1) [2]:

$$v = v_m + v_t \tag{1}$$

In the turbulence domain; the turbulence spectrum observed by a rotating blade element differs from that corresponding to at a fixed point, where a part of the kinetic energy moves toward higher frequencies and will concentrate around integers of rotation frequency [3]. In the stationary domain, the wind speed changes over the rotor disc due to wind shear and tower shadow effects. Wind shear is the

IntechOpen

change in the mean wind speed with the height due to the skin friction of the earth. Tower shadow effect is the reduction of the wind speed when the wind is felt by blades when passing the tower [4].

For an expectation of the realistic WT behavior as response to its control inputs, it is required to have an accurate model of the wind field that takes into consideration all of the above mentioned characteristics. The knowledge of the mean wind speed in a site, which represents the hour to annually changes, is necessary to predict if a wind energy installation is economically profitable. The turbulence component has low impacts on the annual energy produced, while it has significant impacts on the WT dynamics and thus on the power quality produced. Generally, two wind speed profiles including the two components are often employed in literature, an effective wind speed (i.e., a single point wind speed), and a fully field wind speed (i.e., multi points wind speed over the rotor area). The effective wind speed design is based on a mathematical model, while the fully field wind speed is derived from specialized software tools such as TurbSim [5]. The effective wind speed is sufficient in the view of maximum power *capture point*, while the full field wind profile is required to simulate the effects of the aerodynamic conditions on the WT structures.

2. WECS modeling

WECS includes various multidisciplinary subsystems, which can be classified as aerodynamic, structural and electrical. The aerodynamic subsystem represents the aerodynamic model of the WT. The structural subsystems include blades, tower and drive train models. The electrical subsystems include the generator, the back-to-back converter and the system control models. In order to model the dynamic behavior of the overall system, models of the different components need to be derived and coupled consecutively.

2.1 Aerodynamic modeling

The WT aerodynamic subsystem is often modeled by using blade element momentum (BEM) theory [6]. BEM is a *quasi-steady* method, i.e., the assumed local flow conditions and the resulting aerodynamic forces depend only on the current conditions, and would change immediately with any changes in the wind field or blade movement [4]. Under the wind effects, the WT is subjected to surface forces which vary over the rotor area. By applying this theory, the aerodynamic forces acting on each element of the blades can be calculated based on the section of that element. Drag and lift forces, as mentioned above, are the main aerodynamic forces acting on the blades structures, where the lift and the drag are the perpendicular and parallel forces to the incoming flow direction respectively, as illustrated in **Figure 1**. The lift (F_L) and the drag (F_D) forces acting on each blade section based on the local *resultant air velocity* (v), and the lift (c_l) and drag (c_d) coefficients are given by Eqs. (2) and (3), respectively:

$$F_L = \frac{1}{2} c_l(\alpha, R_e) \rho \mathrm{bc} v^2 \tag{2}$$

$$F_D = \frac{1}{2} c_d(\alpha, R_e) \rho \mathrm{bc} v^2 \tag{3}$$

where α, R_e, b and c are local angle of attack, Reynolds number, Blade element length and chord respectively. The lift and drag coefficients are function of the local angle of attack and the Reynolds number. The angle of attack is function of the free

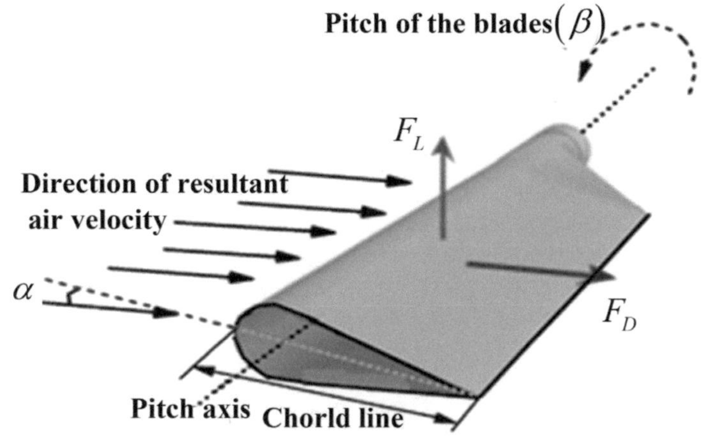

Figure 1.
Lift and drag forces on a blade.

stream wind speed, the rotational speed of the WT, and the local induction factor. The induction factor is the factor by which the free stream wind speed has been slowed down when it reaches the rotor plane. These coefficients are determined either by using wind tunnel tests or computational fluid dynamic simulations [7].

Due to the quadratic influence of the wind speed and the dependence of the lift and drag coefficients, the aerodynamic model of the WT is very complex and highly nonlinear. The lift and drag forces could be used to calculate aerodynamic torque and thrust force on the rotor, by integration from blade root to tip [7].

The aerodynamic torque and the thrust force are given by Eqs. (4) and (5), respectively.

$$T_a = \frac{1}{2}\rho\pi R^3 C_q(\lambda, \beta)v^2 \tag{4}$$

$$F = \frac{1}{2}\pi R^2 C_t(\lambda, \beta)v^2 \tag{5}$$

where C_q and C_t are torque and thrust coefficients. λ and β are tip speed ratio and pitch angle respectively. C_q and C_t can be calculated by the BEM theory, approximated by a mathematical model for simplicity and computational cost or provided by the WT manufacturer in the form of a look-up table [8]. The power coefficient is a nonlinear function of λ and β. Its value depends on the aerodynamic parameters of the turbine as well as on the metrological characteristic of the site. Theoretically this coefficient may take the value of 0.59, which is called the Betz limit, but in reality it is limited between 0.4 and 0.5 [9, 10].

For a detailed modeling, often the aerodynamic model is estimated by specialized aero-elastic simulation tools such as AeroDyn, based on the BEM theory. The outputs from the aerodynamic model are the aerodynamic thrust force and torque, and the distributed forces along each blade. These forces are considered as external forces acting on the structural components of the turbine.

2.2 Structural modeling

The structural model of the WT is a combination of flexible and rigid bodies. It includes tower, blades, drive train, nacelle, main bearing and hub. The choice of the modeling type depends strongly to the dynamics required to capture by the

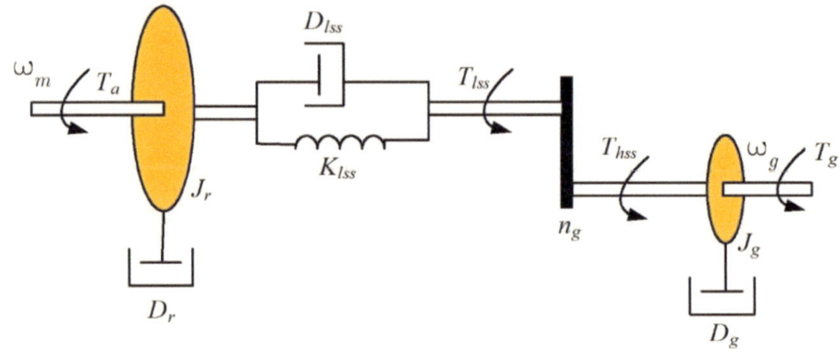

Figure 2.
Two mass model of the drive train.

model. Generally, when interesting to the WT structural dynamics often tower, blades and drive train are modeled as flexible bodies whereas the others are modeled as rigid. In contrast, when interesting to the electrical dynamics the structural dynamics can be simplified and *lumped* into one rigid mass.

2.2.1 Blades

For computational cost simplicity and modal reduction, the flexible bodies motions are assumed to be a superposition of the first eigenmodes. Usually the first and second modes are considered in simulation. This is justified by the fact, that high modes being difficult to detect in actual measurements and having negligible energy content. Therefore, each blade will be modeled by two degree of *freedoms* (DOFs) for flap-wise and one for edge-wise deflection.

2.2.2 Drive train

The drive train allows to convert the high torque with low speed (T_{lss}) to a low torque with a high speed (T_{hss}), by a set of differential equations. The dominant inertias in the WT drive train are: the rotor and generator inertias. The drive train is modeled in this *chapter* by two models: two mass and one mass. In the two mass model, the rotor and generator inertias are separated by a spring, defined by stiffness and damping coefficients, as illustrated by **Figure 2**. In the one mass model, the rotor and the generator inertias are lumped into one mass.

2.2.2.1 Two mass model

In this model, the inertias of the low speed shaft (LSS), the gearbox and the high speed shaft (HSS) are neglected compared to the inertia of the generator and the rotor. The dominant masses in this model are connected by a flexible shaft and characterized by an equivalent torsional stiffness (K_{lss}) and damping factor (D_{lss}), which would be referred either on the LSS or the HSS. The model of the drive train in this *chapter* is referred to the LSS, which is described by the following differential equations:

$$\begin{cases} J_r\dot{\omega}_m = T_a - T_{lss} - D_m\omega_m \\ J_g\dot{\omega}_g = T_{lss} - n_g T_{hss} - D_g\omega_g \\ T_{lss} = K_{lss}\left(\theta_m - \dfrac{\theta_g}{n_g}\right) + D_{lss}\left(\omega_m - \dfrac{\omega_g}{n_g}\right) \end{cases} \quad (6)$$

where ω_m and ω_g are the rotor and the generator speed, θ_m and θ_g are the angular position of the LSS and the HSS side, J_r is the rotor inertia, J_g is the generator inertia referred to the LSS. n_g is the gearbox ratio, which is considered ideal and T_{hss} is the high shaft speed torque. D_m and D_g are the rotor and generator friction coefficient, respectively. T_g is the generator torque.

2.2.2.2 One mass model

As mentioned above, the turbine and generator inertias are lumped into one inertia in the one mass model. The one mass model referred to the LSS, is described by the following differential equation:

$$J_t\dot{\omega}_m = T_a - T_g - D_t\omega_m \tag{7}$$

where the inertia (J_t) and the friction (D_t) coefficient of the lumped mass are expressed by:

$$\begin{cases} J_t = J_r + n_g^2 J_g \\ D_t = D_m + n_g^2 D_g \end{cases}$$

2.3 State space model of the WT

The aero-elastic equation of motion of a WT can be expressed as:

$$M(q, u, t)\ddot{q} + f(q, \dot{q}, u, u_d, t) = 0 \tag{8}$$

where M represents the mass matrix containing inertia and mass components, and f is the nonlinear forcing function vector that includes the stiffness and damping effects. q, $\dot{q}$ and $\ddot{q}$ are the enabled DOFs displacements, velocities and accelerations. u is the control input which is the collective pitch angle, while u_d is the disturbance input. LTI (linear time invariant) model of the WT in the state space representation can be obtained by linearization of the nonlinear model by using aero-elastic software. Fatigue, aerodynamic, structural and turbulence software (FAST) linearizes numerically the equation of motion (Eq. (8)) by perturbing each variable around its respective operating point (OP) [11]. The linearization in FAST is realized in two steps. Firstly, a steady state OP of the enabled DOFs is calculated. Secondly, a numerical linearization about the resulting steady state OP is performed to form periodic matrices of the linear model. The periodic model of the WT in the state space representation, around an OP defined by the triplet $(\overline{\beta}, \overline{v}, \overline{\omega}_g)$ where $\overline{\beta}$, $\overline{\beta}$ and $\overline{\omega}_g$ the optimum value of pitch angle, wind speed and generator speed respectively, is given by:

$$\begin{cases} \dot{x}_m = A_m(\theta_m)x_m + B_m(\theta_m)u_m + B_{dm}(\theta_m)u_{dm} \\ y_m = C_m(\theta_m)x_m + D_m(\theta_m)u_m + D_{dm}(\theta_m)u_{dm} \end{cases} \tag{9}$$

where A_m is the state matrix, B_m is the control matrix, B_{dm} is the disturbance matrix and C_m is the output matrix. The state vector is given by: $x_m = [\Delta q \;\; \Delta\dot{q}]^T$, where Δ represents the perturbation of a parameter from its optimal value. It should be noted that the matrices of the periodic state space model depend on the rotor azimuth position (θ_m). Therefore, it is important to mention that the dynamics of the model of Eq. (9) are expressed in a mixed reference coordinates, as denotes the subscript m, i.e., rotating and fixed references coordinates. In fact, the dynamics of the blades are expressed with respect to rotating reference coordinate, while those of the nacelle and tower are expressed with respect to a fixed reference coordinate. To derive a LTI model from the periodic linear model, it is required to apply a multi-blade coordinate (MBC) transformation [12]. The transformation provides a unifying reference frame for analysis and control design, where all DOFs are expressed with respect to the same reference coordinate. Employing this transformation, the model of Eq. (9) can be transformed to a periodic model expressed in a fixed reference coordinate system:

$$\begin{cases} \dot{x}_{FR} = A_{FR}(\theta_m)x_{FR} + B_{FR}(\theta_m)u_{FR} + B_{dFR}(\theta_m)u_{dFR} \\ y_{FR} = C_{FR}(\theta_m)x_{FR} + D_{FR}(\theta_m)u_{FR} + D_{dFR}(\theta_m)u_{dFR} \end{cases} \tag{10}$$

The subscript *FR* signifies that the model is expressed in the fixed reference coordinate. The states, control inputs and outputs measurement of the model in the rotating reference coordinate are transformed to those corresponding in the fixed reference coordinate using the matrix transformation ($T_c(\theta_m)$), i.e., $x_m = T_c(\theta_m)x_{FR}$, $u_m = T_c(\theta_m)u_{FR}$ and $y_m = T_c(\theta_m)y_{FR}$. The disturbance input (u_{dm}) is not transformed because it is already expressed in the fixed reference coordinate.

$$T_c(\theta_m) = \begin{bmatrix} 1 & \cos(\theta_m) & \sin(\theta_m) \\ 1 & \cos\left(\theta_m + \frac{2\pi}{3}\right) & \sin\left(\theta_m + \frac{2\pi}{3}\right) \\ 1 & \cos\left(\theta_m + \frac{4\pi}{3}\right) & \sin\left(\theta_m + \frac{4\pi}{3}\right) \end{bmatrix} \tag{11}$$

The WT averaged periodic model over the rotor rotational period (θ_m), which is the LTI model, is expressed as:

$$\begin{cases} \dot{x} = Ax + Bu + B_d u_d \\ y = Cx + Du + D_d u_d \end{cases} \tag{12}$$

3. Actuators

The actuators aim to control WECS by receiving the desired signals from the control system and providing the real signals. They include generator, converters, pitch and yaw. The generator actuator receives a desired electromagnetic torque from the torque controller and provides a real electromagnetic torque to the drive train. The converters actuator aims to produce a current in the generator that produces a real torque that matches the desired value, by means of its control system. Generally, the electrical dynamics are much faster compared to the WT structural dynamics. Thus, they can be represented by a small communication delay between the torque demanded by the generator and the actual air-gap torque acting on the mechanical system when the structural dynamics are required. Such delay can be represented by a first order system with a small time constant. The pitch actuator receives the desired pitch angle for each blade from the pitch control, and provides the realistic pitch angle for the blade. The yaw mechanism rotates the nacelle around the vertical tower axis to place the turbine directly into the wind for a maximum power generation. Under high-speed winds, the yaw mechanism turns the blades 90° from the direction of the wind to reduce stress on internal

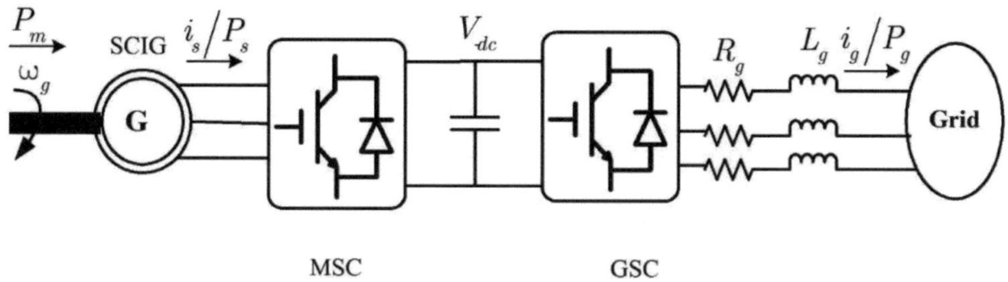

Figure 3.
SCIG interfaced to utility grid.

components and avoid over-speed conditions. The yaw actuator requires the wind inflow direction and nacelle direction as inputs, and provides the yaw angle.

As mentioned above, the pitch angle is used as control input in the full load region, while the generator torque is used as control input in the partial and the full load regions. The yaw actuator is not activated where the WT is aligned with the wind direction, to allow a maximum power capture from the wind.

3.1 Generator modeling

The generator chosen in this manuscript is squirrel *cage* induction generator (SCIG) because it is cheap, simple in construction, easy for maintenance, and easily replaceable. SCIG consists of a set of windings on the stator and rotor. The stator and rotor have three windings corresponding to the three phase system, which can be coupled either in star or delta. The stator phases of the machine are connected to the power grid through a back-to-back converter, whereas those of the rotor are short-circuited by a conductive ring at each winding side, as illustrates **Figure 3**.

3.1.1 Energy generation theory

According to Ferraris theorem, a rotating magnetic field in the gap of the machine having a sliding speed with respect to stator ($\Omega = \frac{\omega_s}{p}$) is created [13]. The rotor turns, with respect to the stator, at the mechanical speed of $\left(\omega_r = \frac{d\theta}{dt}\right)$, where θ is the angle between the stator and the rotor windings. The relative speed between the rotating field and the rotor induces a variable voltage (e_{ind}) in each closed loop of the rotor conductors, where its magnitude linked to the stator flux (Φ_s) by Faraday's law [5]:

$$\frac{d\Phi_s}{dt} = e_{ind} - R_c i_{ind} \tag{13}$$

with i_{ind} is the induced current in the conductor, and R_c is the resistance of the conductor. The stator flux linkage is defined as the flux linkage of a single turn (ϕ) multiplied by the number of turns (N) in the coils:

$$\Phi_s = N\phi$$

3.1.2 SCIG modeling assumptions

Thereafter some assumptions related to modeling of the SCIG are giving in the following of this manuscript [14]:

- The three *phase* system is symmetric and balanced with constant gap.

- The iron saturation is discarded.
- The magnetic flux density is assumed radial to the gap.
- All kind of losses in the iron are neglected.
- Both windings of the machine are sinusoidal distributed, whose axes are displaced by 120°, which produce a sinusoidal magnetic field distribution in the gap.
- The coefficients of the inductances are fixed and the coefficients of the mutual inductances depend to the windings position.

3.1.3 abc and dq0 reference frames

The quadratic-direct-zero ($dq0$) (see **Figure 4**), proposed by Park, is a reference rotating frame, for modeling and analysis of the electrical systems [15]. It allows to remove the dependency of certain parameters like inductances, and to simplify the modeling process. Generally, the electrical systems are modeled in phases reference frame. For control and analysis purposes, it is required to have the $dq0$ model of the system. The passage from a reference frame to the other is performed by means of the Park transformation matrix. By application of this transformation, the three ac (a, b and c) quantities of the generator can be reduced to two DC (d and q) components and a homopolar component.

Assuming that the dq frame is oriented by an angle (θ) from the abc reference frame, the Park transformation matrix to convert the abc quantities of the SCIG to the $dq0$ quantities is given by:

$$T(\theta) = \frac{2}{3}\begin{bmatrix} \cos(\theta) & \cos\left(\theta - \frac{2\pi}{3}\right) & \cos\left(\theta + \frac{2\pi}{3}\right) \\ \sin(\theta) & \sin\left(\theta - \frac{2\pi}{3}\right) & \sin\left(\theta + \frac{2\pi}{3}\right) \\ \frac{1}{2} & \frac{1}{2} & \frac{1}{2} \end{bmatrix} \tag{14}$$

Therefore, each quantity can be transformed from the abc to the $dq0$ reference frame by the following expression:

$$\chi^{dq0} = T(\theta)\chi^{abc} \tag{15}$$

where χ^{dq0} and χ^{abc} are the representation of a machine parameter in the $dq0$ and abc reference frames, respectively. χ may represent voltage, current or flux of the machine.

The inverse transformation of each quantity can be recovered by means of this expression:

$$\chi^{abc} = T^{-1}(\theta)\chi^{dq0} \tag{16}$$

where

$$T^{-1}(\theta) = \begin{bmatrix} \cos(\theta) & \sin(\theta) & 1 \\ \cos\left(\theta - \frac{2\pi}{3}\right) & \sin\left(\theta - \frac{2\pi}{3}\right) & 1 \\ \cos\left(\theta + \frac{2\pi}{3}\right) & \sin\left(\theta + \frac{2\pi}{3}\right) & 1 \end{bmatrix} \tag{17}$$

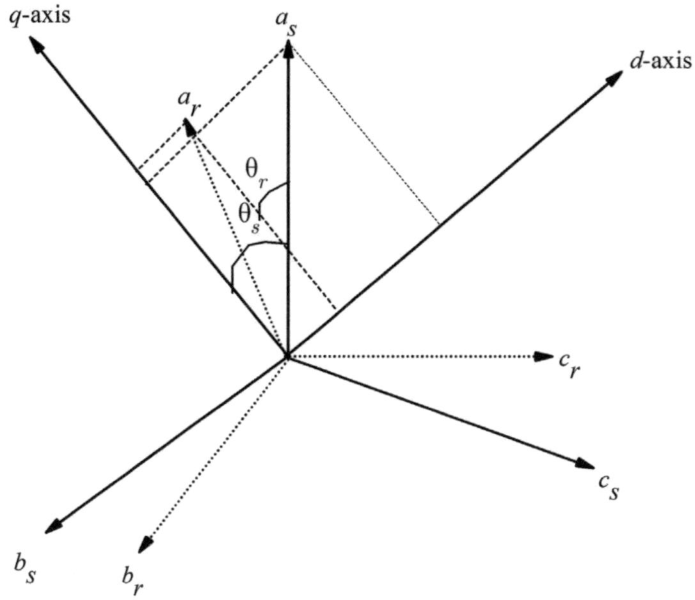

Figure 4.
dq0 reference frame.

3.1.4 SCIG model in the abc reference frame

Applying Faraday law's, under the above mentioned hypotheses and by using the induction motor's illustration, the dynamic model of an induction machine in the *abc* reference frame is described by the following differential equation [14, 16, 17]:

$$\begin{Bmatrix} v_s^{abc} \\ v_r^{abc} \end{Bmatrix} = \begin{bmatrix} r_s^{abc} & 0 \\ 0 & r_r^{abc} \end{bmatrix} \begin{Bmatrix} i_s^{abc} \\ i_r^{abc} \end{Bmatrix} + \frac{d}{dt} \begin{Bmatrix} \Phi_s^{abc} \\ \Phi_r^{abc} \end{Bmatrix} \tag{18}$$

The voltage, current and flux in the *abc* reference frame are expressed as:

$$v_s^{abc} = \begin{Bmatrix} v_{sa} \\ v_{sb} \\ v_{sc} \end{Bmatrix}, v_r^{abc} = \begin{Bmatrix} v_{ra} \\ v_{rb} \\ v_{rc} \end{Bmatrix}, i_s^{abc} = \begin{Bmatrix} i_{sa} \\ i_{sb} \\ i_{sc} \end{Bmatrix}, i_r^{abc} = \begin{Bmatrix} i_{ra} \\ i_{rb} \\ i_{rc} \end{Bmatrix}, \Phi_s^{abc} = \begin{Bmatrix} \Phi_{sa} \\ \Phi_{sb} \\ \Phi_{sc} \end{Bmatrix} \text{ and}$$

$$\Phi_r^{abc} = \begin{Bmatrix} \Phi_{ra} \\ \Phi_{rb} \\ \Phi_{rc} \end{Bmatrix}$$

where Φ_s^{abc} and Φ_r^{abc} are the stator and rotor linkage fluxes in the *abc* reference frame, which are given by:

$$\begin{Bmatrix} \Phi_s^{abc} \\ \Phi_r^{abc} \end{Bmatrix} = \begin{bmatrix} L_{ss}^{abc} & L_{sr}^{abc} \\ L_{rs}^{abc} & L_{rr}^{abc} \end{bmatrix} \begin{Bmatrix} i_s^{abc} \\ i_r^{abc} \end{Bmatrix} \tag{19}$$

By which r_s^{abc}, L_{ss}^{abc}, r_r^{abc} and L_{rr}^{abc} are the resistances and inductances of the rotor and stator in the *abc* reference frame, which are expressed as:

$$r_s^{abc} = \begin{bmatrix} r_s & 0 & 0 \\ 0 & r_s & 0 \\ 0 & 0 & r_s \end{bmatrix}, r_r^{abc} = \begin{bmatrix} r_r & 0 & 0 \\ 0 & r_r & 0 \\ 0 & 0 & r_r \end{bmatrix},$$

$$L_{ss}^{abc} = -\begin{bmatrix} L_{s+}L_m & \frac{-L_m}{2} & \frac{-L_m}{2} \\ \frac{-L_m}{2} & L_{s+}L_m & \frac{-L_m}{2} \\ \frac{-L_m}{2} & \frac{-L_m}{2} & L_{s+}L_m \end{bmatrix} \text{ and } L_{rr}^{abc} = \begin{bmatrix} L_{r+}L_m & \frac{-L_m}{2} & \frac{-L_m}{2} \\ \frac{-L_m}{2} & L_{r+}L_m & \frac{-L_m}{2} \\ \frac{-L_m}{2} & \frac{-L_m}{2} & L_{r+}L_m \end{bmatrix}.$$

L_s and L_r are the inductance value of stator and rotor windings in each phase, respectively. L_m can represent the stator magnetizing inductance, the stator-rotor mutual inductance, and the rotor magnetizing inductance all referred to the stator side.

$L_{sr}^{abc}$and L_{rs}^{abc} are the mutual inductances between the stator and rotor phases for the rotor and stator respectively, which are expressed as:

$$L_{sr}^{abc} = L_m \begin{bmatrix} cos\,(\theta_r) & cos\left(\theta_r + \frac{2\pi}{3}\right) & cos\left(\theta_r - \frac{2\pi}{3}\right) \\ cos\left(\theta_r - \frac{2\pi}{3}\right) & cos\,(\theta_r) & cos\left(\theta_r + \frac{2\pi}{3}\right) \\ cos\left(\theta_r + \frac{2\pi}{3}\right) & cos\left(\theta_r - \frac{2\pi}{3}\right) & cos\,(\theta_r) \end{bmatrix},$$

$$L_{rs}^{abc} = L_m \begin{bmatrix} cos\,(\theta_r) & cos\left(\theta_r - \frac{2\pi}{3}\right) & cos\left(\theta_r + \frac{2\pi}{3}\right) \\ cos\left(\theta_r + \frac{2\pi}{3}\right) & cos\,(\theta_r) & cos\left(\theta_r - \frac{2\pi}{3}\right) \\ cos\left(\theta_r - \frac{2\pi}{3}\right) & cos\left(\theta_r + \frac{2\pi}{3}\right) & cos\,(\theta_r) \end{bmatrix}.$$

r_s and r_r are resistance value of stator and rotor windings in each phase, respectively. The induction machine coupling inductance matrix is defined by:

$$L(\theta_r) = \begin{bmatrix} L_{ss}^{abc} & L_{sr}^{abc} \\ L_{rr}^{abc} & L_{rs}^{abc} \end{bmatrix}$$

The rotor quantities referred to the stator are given by: $v_r^{abc} = (N_s/N_r)v_r^{abc}$, $i_r^{abc} = (N_s/N_r)i_r^{abc}$, $r_r^{abc} = (N_s/N_r)^2 r_r^{abc}$, $L_{sr}^{abc} = (N_s/N_r)^2 L_{sr}^{abc}$, $L_{rs}^{abc} = (N_s/N_r)^2 L_{rs}^{abc}$ and $L_{rr}^{abc} = (N_s/N_r)^2 L_{rr}^{abc}$. Where where N_s and N_r are the number of turns per phase in the stator and rotor windings, respectively.

The angular displacement of the rotor, with respect to the stator, and the stator are given by Eqs. (20) and (21), respectively:

$$\theta_r(t) = \int \omega_r(t)dt + \theta_r(0) \tag{20}$$

$$\theta_s(t) = \int \omega_s(t)dt + \theta_s(0) \tag{21}$$

where $\theta_r(0)$ and $\theta_s(0)$ are the initial position of the rotor and stator at t = 0, respectively. The rotor speed is linked to the mechanical speed of the generator ω_g by:

$$\omega_r = p\omega_g \tag{22}$$

The generator speed is linked to the turbine rotational speed (ω_m) by:

$$\omega_g = n_g\omega_m \tag{23}$$

3.1.5 SCIG model in dq0 reference frame

The SCIG model in the $dq0$ reference frame is obtained by applying the transformation matrices $T(\theta_s)$ and $T(\theta_s - \theta_r)$ on the *abc* stator and rotor parameters, respectively. Applying these transformations on the Eqs. (14) and (17) we can write

$$\begin{cases} v_s^{dq0} = T(\theta_s) r_s^{dq0} i_s^{dq0} T^{-1}(\theta_s) + T(\theta_s)\dfrac{d}{dt}\{T^{-1}(\theta_s)\Phi_s^{dq0}\} \\ v_r^{dq0} = T(\theta_s - \theta_r) r_r^{dq0} i_r^{dq0} T^{-1}(\theta_s - \theta_r) + T(\theta_s - \theta_r)\dfrac{d}{dt}\{T^{-1}(\theta_s - \theta_r)\Phi_r^{dq0}\} \end{cases} \tag{24}$$

After simplifications, the stator and rotor voltages in the $dq0$ reference frame are given by:

$$\begin{cases} v_s^{dq0} = r_s^{dq0} i_s^{dq0} + \dfrac{d}{dt}\Phi_s^{dq0} + \Omega_s\Phi_s^{dq0} \\ v_r^{dq0} = r_r^{dq0} i_r^{dq0} + \dfrac{d}{dt}\Phi_r^{dq0} + \Omega_r\Phi_r^{dq0} \end{cases}. \tag{25}$$

where the voltage, the current and the flux in the $dq0$ are expressed as:

$$v_s^{dq0} = \begin{Bmatrix} v_{sd} \\ v_{sq} \\ v_{s0} \end{Bmatrix},\ v_r^{dq0} = \begin{Bmatrix} v_{rd} \\ v_{rq} \\ v_{r0} \end{Bmatrix},\ i_s^{dq0} = \begin{Bmatrix} i_{sd} \\ i_{sq} \\ i_{s0} \end{Bmatrix},\ i_r^{dq0} = \begin{Bmatrix} i_{rd} \\ i_{rq} \\ i_{r0} \end{Bmatrix},$$

$$\Phi_s^{dq0} = \begin{Bmatrix} \Phi_{sd} \\ \Phi_{sq} \\ \Phi_{s0} \end{Bmatrix} \text{ and } \Phi_r^{dq0} = \begin{Bmatrix} \Phi_{rd} \\ \Phi_{rq} \\ \Phi_{r0} \end{Bmatrix}$$

By which Ω_s and Ω_r are the stator and rotor speed transformation matrices, which are given by: $\Omega_s = \begin{bmatrix} 0 & \omega_s & 0 \\ -\omega_s & 0 & 0 \\ 0 & 0 & 0 \end{bmatrix}$ and $\Omega_r = \begin{bmatrix} 0 & (\omega_s - \omega_r) & 0 \\ -(\omega_s - \omega_r) & 0 & 0 \\ 0 & 0 & 0 \end{bmatrix}$, respectively.

The stator and rotor resistances in the *abc* and $dq0$ are equal, i.e., $r_s^{dq0} = r_s^{abc}$ and $r_r^{dq0} = r_r^{abc}$.

Applying the previous transformations (Eqs. (14) and (17)), we obtain:

$$\begin{cases} T^{-1}(\theta_s)\Phi_s^{dq0} = L_{ss}^{dq0}T^{-1}(\theta_s)\mathrm{i}_s^{dq0} + L_{sr}^{dq0}T^{-1}(\theta_s - \theta_r)\mathrm{i}_r^{dq0} \\ T^{-1}(\theta_s - \theta_r)\Phi_r^{dq0} = L_{rs}^{dq0}T^{-1}(\theta_s - \theta_r)\mathrm{i}_s^{dq0} + L_{rr}^{dq0}T^{-1}(\theta_s - \theta_r)\mathrm{i}_r^{dq0} \end{cases} \tag{26}$$

After simplifications, the stator and rotor fluxes of the SCIG in the $dq0$ reference frame are given as:

$$\begin{cases} \Phi_s^{dq0} = L_{ss}^{dq0}\mathrm{i}_s^{dq0} + L_{sr}^{dq0}\mathrm{i}_r^{dq0} \\ \Phi_r^{dq0} = L_{rs}^{dq0}\mathrm{i}_s^{dq0} + L_{rr}^{dq0}\mathrm{i}_r^{dq0} \end{cases} \tag{27}$$

where $L_{ss}^{dq0} = \begin{bmatrix} L_s + \frac{3}{2}L_m & 0 & 0 \\ 0 & L_s + \frac{3}{2}L_m & 0 \\ 0 & 0 & L_{ls} \end{bmatrix}$,

$$L_{rs}^{dq0} = L_{sr}^{dq0} = \begin{bmatrix} \frac{3}{2}L_m & 0 & 0 \\ 0 & \frac{3}{2}L_m & 0 \\ 0 & 0 & \frac{3}{2}L_m \end{bmatrix} \text{ and } L_{rr}^{dq0} = \begin{bmatrix} L_r + \frac{3}{2}L_m & 0 & 0 \\ 0 & L_r + \frac{3}{2}L_m & 0 \\ 0 & 0 & L_r \end{bmatrix}$$

Since the rotor windings are short circuited, the *abc* rotor voltage components are zero, $\{v_r^{abc} = 0\}$. Moreover, as the electric system is supposed equilibrate, the electrical parameters of the model in the $dq0$ reference frame are reduced only to two components (d and q), i.e., the homopolar components equal zero ($v_{0s} = v_{0r} = i_{0s} = i_{0r} = \Phi_{0s} = \Phi_{0r}$). Therefore, the dynamic model of the SCIG machine in the dq reference frame is expressed as:

$$\begin{cases} v_{sq} = r_s i_{sq} + \frac{d\Phi_{sq}}{dt} + \omega_s \Phi_{sd} \\ v_{sd} = r_s i_{sd} + \frac{d\Phi_{sd}}{dt} - \omega_s \Phi_{sq} \\ v_{rq} = r_r i_{rq} + \frac{d\Phi_{rq}}{dt} + (\omega_s - \omega_r)\Phi_{rd} = 0 \\ v_{rd} = r_r i_{rd} + \frac{d\Phi_{rd}}{dt} - (\omega_s - \omega_r)\Phi_{rq} = 0 \end{cases} \tag{28}$$

Figure 5 represents the electrical scheme of the SCIG in the dq reference frame based on the electrical model of Eq. (28).

After simplifications, the electromagnetic flux in the dq reference frame of the SCIG is given by the following equation:

$$\begin{cases} \Phi_{sq} = L_s i_{sq} + L_M i_{rq} \\ \Phi_{sd} = L_s i_{sd} + L_M i_{rd} \\ \Phi_{rq} = L_r i_{rq} + L_M i_{sq} \\ \Phi_{rd} = L_r i_{rd} + L_M i_{sd} \end{cases} \tag{29}$$

$$\text{with} \begin{cases} L_s = L_{ls} + L_M \\ L_r = L_{lr} + L_M \\ L_M = \frac{3}{2}L_m \end{cases}.$$

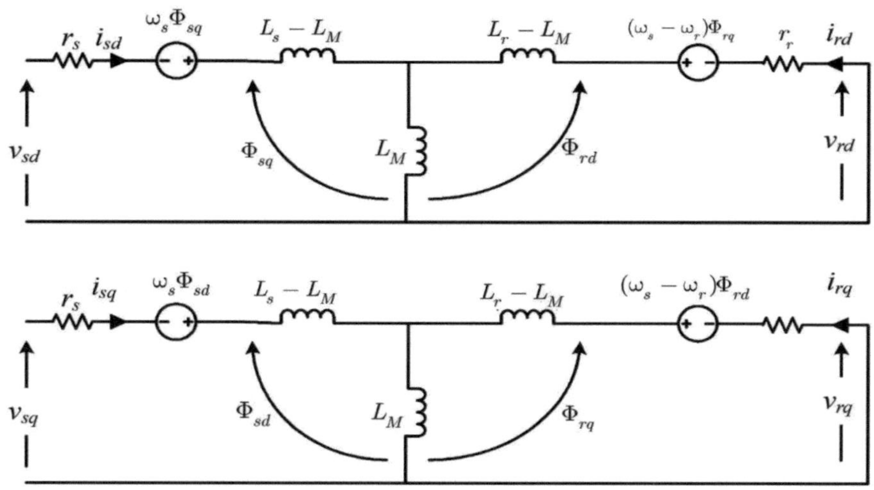

Figure 5.
Electrical scheme of the SCIG in the dq reference frame.

3.1.6 State space model of the SCIG machine in the dq0 reference frame

Combining Eqs. (28) and (29), the model of the SCIG in the state space representation can be derived, where $x_n = \left[i_{sq}, i_{sd}, i_{rq}, i_{rd}\right]^T$ is the state vector, $u_n = \left[v_{sq}, v_{sd}\right]^T$ is the control input. The state (A) and control input (B) matrices of the model are given by:

$$A = \begin{bmatrix} -\dfrac{r_s}{\sigma L_s} & \omega_s + \dfrac{p\omega_g L_m^2}{\sigma L_s L_r} & \dfrac{L_m L_r}{\sigma L_s L_r} & \dfrac{p\omega_g L_m}{\sigma L_s} \\ -\left(\omega_s + \dfrac{p\omega_g L_m^2}{\sigma L_s L_r}\right) & -\dfrac{r_s}{\sigma L_s} & -\dfrac{p\omega_g L_m}{\sigma L_s} & \dfrac{L_m L_r}{\sigma L_s L_r} \\ \dfrac{L_m r_s}{\sigma L_s L_r} & -\dfrac{p\omega_g L_m}{\sigma L_r} & -\dfrac{r_r}{\sigma L_r} & \omega_s - \dfrac{p\omega_g}{\sigma} \\ \dfrac{p\omega_g L_m}{\sigma L_r} & \dfrac{L_m L_r}{\sigma L_s L_r} & \dfrac{p\omega_g}{\sigma} - \omega_s & -\dfrac{r_r}{\sigma L_r} \end{bmatrix},$$

$$B = \begin{bmatrix} \dfrac{1}{\sigma L_s} & 0 \\ 0 & \dfrac{1}{\sigma L_s} \\ -\dfrac{L_m}{\sigma L_s L_r} & 0 \\ 0 & -\dfrac{L_m}{\sigma L_s L_r} \end{bmatrix}$$

where $\sigma = 1 - L_M^2 / L_s L_r$.

3.1.7 Active power and electrical torque in the dq0 reference frame

The electromagnetic torque T_g developed by the machine depends to the instantaneous current circulated in each of the six windings and the angle between the stator and rotor windings θ. Its expression in the *abc* reference frame is given as [17]:

$$T_g = \frac{p}{2} \begin{Bmatrix} i_s^{abc} \\ i_r^{abc} \end{Bmatrix}^T [L(\theta)] \begin{Bmatrix} i_s^{abc} \\ i_r^{abc} \end{Bmatrix} \tag{30}$$

Its model in the $dq0$ reference frame is given by [14]:

$$T_g = \frac{3}{2} p \left(\Phi_{rd} i_{rq} - \Phi_{rq} i_{rd} \right) \tag{31}$$

The active power (P_s) yield by the SCIG is given as the product of the electromagnetic torque (T_g) and the generator speed (ω_g):

$$P_s = T_g \omega_g \tag{32}$$

Its expression in the $dq0$ reference frame, function of voltages and currents stator, is given by:

$$P_s = \frac{3}{2} \left(v_{sd} i_{sd} + v_{sq} i_{sq} \right) \tag{33}$$

3.2 Converters modeling

The back to back converter interfaces the WT and the utility grid. It consists of a machine side converter (MSC), which acts as rectifier, and GSC, which acts as inverter, connected by a DC link as illustrated in **Figure 6**. The DC link allows an optimum injection of the generated energy to the grid, and a frequency decoupling between the machine and the grid. The structure of the converters is bi-directional power flow, i.e., the direction of the power flow can be reversed at any time. The converter outlined in this chapter is a voltage source converter (VSC), which is a switching IGBT-diode device. Its switching model describes the steady state and the dynamic behavior, including slow transients and high frequency components of voltages and currents. The high frequency of the components are not desirable for control and analysis purposes, thus it is common to use the average value of the variables rather than the instantaneous value. Therefore, only the fundamental ac dynamics which are considered for control design and analysis [5].

The main objective associated to the rectifier is to control the WT variable speed operation, i.e., to implement the MPPT (maximum power point tracking) strategy, whereas that associated to the inverter is to manage the power flow exchanged

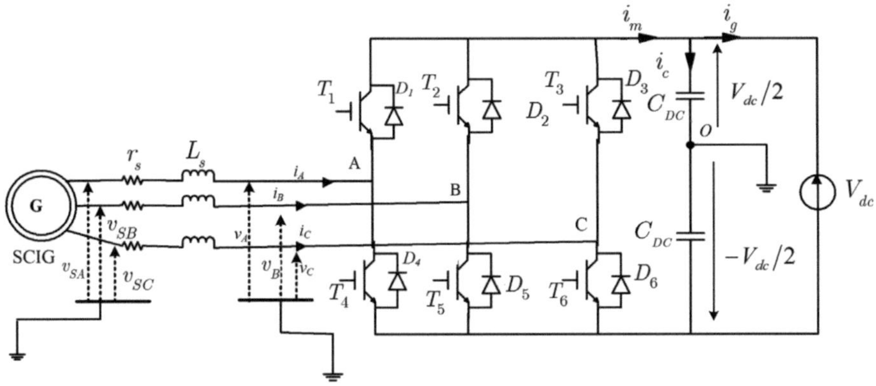

Figure 6.
MSC electrical scheme.

between the machine and the utility grid. As the modeling process of these converters are the same, we will here only to MSC. **Figure 6** illustrates the connecting scheme of the MSC to the DC link. As can be shown, each phase of the converter is composed of a set of two IGBT-diode devices connected to the DC circuit. It has been proven that the efficiency of the converter is around 94–98%, thus it is common practice to represent the back to back converters as ideal switches. Therefore, each set of the IGBT-diode of the phase will be represented by a single switch, where its switches states are complementary. Moreover, for analysis purpose, it is assumed that there is a common point in the DC link that splits the voltage V_{dc} into two voltages of amplitudes $V_{dc}/2$ [17].

3.2.1 MSC modeling assumptions

Thereafter are recalled some assumptions related to the converter modeling [14]:

- The switching of the components is instantaneous.
- The voltage drops at their terminals are negligible.
- The dead times are neglected.
- The load is three-phase balanced and coupled in star with isolated neutral point.
- It is assumed that the IGBTs switch at a frequency at least 10 times of the fundamental frequency of the ac grid, thus the action of this commutation can be represented by the average of the duty cycle [5]. Therefore, the currents and voltages of the VSC are expressed as function of the duty cycle of the IGBT.

3.2.2 MSC model in abc *reference frame*

Applying Kirchhoff's law to circuit given in **Figure 6**, the compound voltages between the three phases at the right side, are:

$$\begin{cases} u_{AB} = u_{AO} - u_{BO} \\ u_{BC} = u_{BO} - u_{CO} \\ u_{CA} = u_{CO} - u_{AO} \end{cases} \tag{34}$$

At the load side, the compound voltages are function of the simple voltages, i.e., the voltages between the phases A, B and C and the point O, are expressed as:

$$\begin{cases} u_{AB} = v_A - v_B \\ u_{BC} = v_B - v_C \\ u_{CA} = v_C - v_A \end{cases} \tag{35}$$

The voltages equations of the VSC can be expressed also as [17]:

$$\begin{cases} u_{AB} - u_{CA} = 2v_A - (v_B + v_C) \\ u_{BC} - u_{AB} = 2v_B - (v_C + v_A) \\ u_{CA} - u_{BC} = 2v_C - (v_A + v_B) \end{cases} \tag{36}$$

Moreover, as the voltages v_A, v_B, v_C form an equilibrium three phase system, the expression of Eq. (36) can be reduced to [5]:

$$\begin{cases} u_{AB} - u_{CA} = 3v_A \\ u_{BC} - u_{AB} = 3v_B \\ u_{CA} - u_{BC} = 3v_C \end{cases} \quad (37)$$

Thus, the simple voltages function of the voltages between the phases and the effective point O, are expressed by:

$$\begin{cases} v_A = \frac{1}{3}(u_{AB} - u_{CA}) \\ v_B = \frac{1}{3}(u_{BC} - u_{AB}) \\ v_C = \frac{1}{3}(u_{CA} - u_{BC}) \end{cases} \quad (38)$$

$$\begin{bmatrix} v_A \\ v_B \\ v_C \end{bmatrix} = \frac{1}{3}\begin{bmatrix} 2 & -1 & -1 \\ -1 & 2 & -1 \\ -1 & -1 & 2 \end{bmatrix}\begin{bmatrix} u_{AO} \\ u_{BO} \\ u_{CO} \end{bmatrix} \quad (39)$$

As mentioned above, each phase is formed by two complementary ideal switches, i.e., if we associate to each phase a binary value of command S_i (i = A, B, C) we have:

S_i = +1, the upper switch is on and the lower switch is off;
S_i = −1, the upper switch is off and the lower switch is on.

Therefore, the voltages between the phases and the effective point O, function of the switch functions and the DC link voltage are derived

$$\begin{bmatrix} u_{AO} \\ u_{BO} \\ u_{CO} \end{bmatrix} = \frac{V_{dc}}{2}\begin{bmatrix} S_A \\ S_B \\ S_C \end{bmatrix} \quad (40)$$

Combining Eqs. (39) and (40), the simple phase voltages, function of the switching functions and the DC link voltage, are obtained:

$$\begin{bmatrix} v_A \\ v_B \\ v_C \end{bmatrix} = \frac{1}{3}\begin{bmatrix} 2 & -1 & -1 \\ -1 & 2 & -1 \\ -1 & -1 & 2 \end{bmatrix}\frac{V_{dc}}{2}\begin{bmatrix} S_A \\ S_B \\ S_C \end{bmatrix} \quad (41)$$

Finally, the simple voltages of the three phases at the output of the converter are given function of the switch functions. The switching functions are the commands of the converter, where their states will be provided from PWM control strategy.

Moreover, applying Kirchhoff's law at the load side, the average model of the VSC, in the *abc* reference frame is given by:

$$\begin{cases} L_s\frac{di_A}{dt} + r_s i_A = v_{SA} - v_A \\ L_s\frac{di_B}{dt} + r_s i_B = v_{SB} - v_B \cdot \\ L_s\frac{di_C}{dt} + r_s i_C = v_{SC} - v_C \end{cases} \quad (42)$$

where v_{SA}, v_{SB} and v_{SC} are the induced voltages in the stator phases of the generator.

3.2.3 MSC model in the dq0 reference frame

Expressing the voltages and currents of the VSC in the dq reference frame allows controlling independently the electromagnetic torque and rotor flux.

The model of the MSC in the $dq0$ reference frame can be obtained by application of the transformation matrix of Eq. (14) on Eq. (42) [8]:

$$\begin{cases} L_s \dfrac{di_{sd}}{dt} + r_s i_{sd} = v_{sd} - S_d \dfrac{V_{dc}}{2} \\ L_s \dfrac{di_{sq}}{dt} + r_s i_{sq} = v_{sq} - S_q \dfrac{V_{dc}}{2} \end{cases} \quad (43)$$

where S_d and S_q are the d and q switch functions, respectively.

This model will be used to control the SCIG machine by controlling the VSC based on PWM control strategy. Selective harmonic elimination, sinusoidal pulse width modulation (SPWM) and space vector modulation are the commonly found PWM strategies in literature. SPWM is the most used due to its simplicity and effectiveness [5].

3.2.4 SPWM control strategy

SPWM is a multi-pulses-based modulation method that varies the pulse width of the converter output voltage in a sinusoidal manner, following a target reference voltage by comparing a low frequency reference signal (f_m) with a high frequency carrier signal $\left(f_p\right)$, where $f_m \ll f_p$ [5]. The frequency of the reference signal should equal the frequency of the output voltage. For each phase leg, there are two complementary standing signals, one for the lower switch and the other for the upper switch. In fact, when the reference signal exceeds the carrier signal, the upper switch is turned on and the lower switch is off. Otherwise if the reference signal is less that the carrier signal, the lower switch is turned on and the upper switch is turned off.

For a three phase VSC, the three phase reference signals v_{mA}, v_{mB}, and v_{mC} and a high frequency carrier (v_p) are used

$$\begin{cases} v_{mA} = \hat{V}_m \sin\left(2\pi f_m t\right) \\ v_{mB} = \hat{V}_m \sin\left(2\pi f_m t - \dfrac{2\pi}{3}\right) \\ v_{mC} = \hat{V}_m \sin\left(2\pi f_m t + \dfrac{2\pi}{3}\right) \end{cases} \quad (44)$$

$$v_p = \begin{cases} \hat{V}_p\left(-1 + 4\dfrac{t}{T_p}\right) \\ \hat{V}_p\left(3 - 4\dfrac{t}{T_p}\right) \end{cases} \text{if} \begin{cases} t \in \left[0, \dfrac{T_p}{2}\right] \\ t \in \left[\dfrac{T_p}{2}, 0\right] \end{cases}$$

where $\hat{V}_m$ and $\hat{V}_p$ are the voltages amplitude, and T_p is the period of the carrier. The states of the IGBT (S_j) (j = 1,2,3) are derived as follow:

$$S_j = \begin{cases} +1 \\ -1 \end{cases} \text{if} \begin{cases} v_{mi} - v_p \geq 0 \\ v_{mi} - v_p < 0 \end{cases}$$

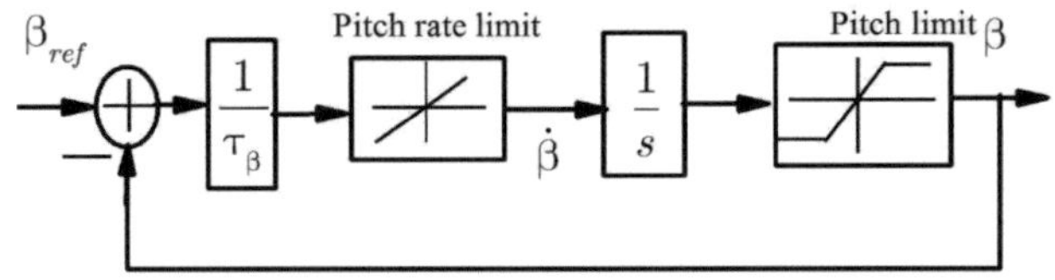

Figure 7.
Pitch actuator scheme.

where the states of the lower switches will be complementary with those of the upper switches. Thus the simple voltages (v_i) at the terminal of the converter are:

$$v_i = \begin{cases} +\dfrac{V_{dc}}{2} \\ -\dfrac{V_{dc}}{2} \end{cases} \text{if} \begin{cases} S_j = +1 \\ S_j = -1 \end{cases}$$

Therefore, the voltages at the output of the converter are a set of rectangular slots, whose width varies as function of the control law of the IGBT, the duration of switch and the time of functioning. Two parameters characterize this control strategy: the modulation index, which is the ratio of f_m and f_p; and the tuning coefficient, which is the ratio of $\hat{V}_m$ and $\hat{V}_p$. For a proper control, these coefficients must be tuned carefully. Various works are interested to WTs control either for maximum capture, power flow exchange as well as power regulation and structural loads minimization. Readers can refer to [18–23].

3.3 Pitch actuator

The pitch actuator system aims to drive the blades to the desired positions received from the pitch control. Due to the large inertia of the blades, there is a significant delay between a blade pitch command and the blade actually arriving in the desired position. The actuator dynamics are largely dominated by the blades and the motor inertias, and the capabilities of the pitch system; while the torque caused by the aerodynamic loading of the blades has only small impacts. Often the pitch actuator model is linear first order system with low pass filter, additional rate limiter acting on the demanded pitch angle $\left(\beta_{ref}\right)$ and angle limiter to set the maximum and the minimum pitch angles. The pitch system can be either hydraulic, electromechanical, or a hybrid of the both. Eq. (45) represents the model of the actuator, where τ_β is the time constant of the model.

$$\frac{\beta}{\beta_{ref}} = \frac{1}{\tau_\beta . s + 1} \tag{45}$$

Figure 7 illustrates the pitch actuator scheme based on the mathematical model of Eq. (45).

4. Conclusions

This chapter presented the different models of the WECS such as wind turbines aerodynamics and structural dynamics, generator and converter actuators. The chapter has outlined the hypotheses and the approaches used in control and modeling such as state space representation, and the *abc* and the *dq* reference frames. The chapter finished by presentation of the SPWM control strategy and the pitch angle actuator.

Conflict of interest

The authors do not have any conflicts of interest to declare.

Nomenclature

abc	three phases system
dq	park phases
θ_s	position angle between the *abc* and the *dq* reference frames
θ_r	position angle between the rotor and the stator windings in the *abc* reference frame
a_r, b_r, c_r	rotor windings axes in the *abc* reference frame
a_s, b_s, c_s	stator windings axes in the *abc* reference frame
P_s	electrical power
T_g	electromagnetic torque
ω_g	generator speed
i_s	stator current
P_g	electrical energy injected to grid
L_g	filter inductance
R_s	filter resistances
i_g	grid current
V_{dc}	DC link voltage
MSC	machine side converter
GSC	grid side converter
VSC	voltage source converter

Author details

Ali El Yaakoubi*, Lahcen Amhaimar and Adel Asselman
Optic and Photonic Team, Abdelmalek Essaadi University, Tetouan, Morocco

*Address all correspondence to: ali.elyaakoubi@gmail.com

References

[1] Van der Hoven I. Power spectrum of horizontal wind speed in the frequency range from 0.0007 to 900 cycles per hour. Journal of Meteorology. 1957;**14**: 160-164. DOI: 10.1175/1520-0469(1957) 014<0160:PSOHWS>2.0.CO;2

[2] Nichita C, Luca D, Dakyo B, Ceanga E. Large band simulation of the wind speed for real time wind turbine simulators. IEEE Transactions on Energy Conversion. 2002;**17**:523-529. DOI: 10.1109/TEC.2002.805216

[3] Bianchi FD, de Battista H, Mantz RJ. Wind Turbine Control Systems: Principles, Modelling and Gain Scheduling Design. London: Springer-Verlag; 2007

[4] Arne K. Extreme and fatigue load reducing control for wind turbines: A model predictive control approach using robust state constraints [thesis]. Berlin: von der Fakultät III—Prozesswissens der Technischen Universität Berlin; 2014

[5] El Yaakoubi A. Wind energy conversion systems control for maximum power capture and excessive loads alleviation [thesis]. Tétouan, Morocco: Abdelmalek Essaadi University; 2018

[6] Hansen MOL. Aerodynamics of Wind Turbines. 3rd ed. New York, NY: Routledge; 2015

[7] Ingram G. Wind turbine blade analysis using the blade element momentum method. version 1.0 [thesis]. UK: School of Engineering, Durham University; 2005

[8] Munteanu I, Bratcu AI, Cutululis N-A, Ceanga E. Optimal Control of Wind Energy Systems: Towards a Global Approach. London: Springer-Verlag; 2008

[9] Chen Z, Guerrero JM, Blaabjerg F. A review of the state of the art of power electronics for wind turbines. IEEE Transactions on Power Electronics. 2009;**24**:1859-1875. DOI: 10.1109/TPEL.2009.2017082

[10] Ragheb M, Ragheb AM. Wind turbines theory—The betz equation and optimal rotor tip speed ratio. In: Carriveau R, editor. Fundamental and Advanced Topics in Wind Power. IntechOpen; 2011. DOI: 10.5772/21398

[11] Jonkman JM, Buhl ML Jr. Fast user's guide-updated August 2005 [Technical Report]. Golden-Colorado: National Renewable Energy Laboratory (NREL); 2005

[12] Bir GS. User's guide to mbc3: Multi-blade coordinate transformation code for 3-bladed wind turbine [Technical Report]. Golden-Colorado: National Renewable Energy Laboratory (NREL); 2010

[13] El Aimani S. Modélisation des différentes technologies d'éoliennes intégrées dans un réseau de moyenne tension [thesis]. Ecole Centrale de Lille; 2004

[14] García D. Modeling and control of squirrel cage induction generator with full power converter applied to windmills [thesis]. Spain: Escola Tècnica Superior d'Enginyeria Industrial de Barcelona—Enginyeria Industrial; 2009

[15] Park RH. Two-reaction theory of synchronous machines generalized method of analysis—Part I. Transactions of the American Institute of Electrical Engineers. 1929;**48**:716-727. DOI: 10.1109/T-AIEE.1929.5055275

[16] Fateh F. Nonlinear control schemes for extremum power seeking and

torsional vibration mitigation in variable speed wind turbine systems [thesis]. USA: Kansas State University; 2015

[17] Olimpo A-L, David C-G, Edgar M-G, Grain A. Offshore Wind Energy Generation: Control, Protection, and Integration to Electrical Systems. Chichester: John Wiley & Sons Ltd; 2017

[18] Yaakoubi AE, Hajri MEH, Attari K, Amhaimar L, Asselman A. Advanced control strategies of SCIG based wind turbines for MPPT. In: Proceedings 3rd International Conference Smart City Applications; New York, NY, USA: ACM; 2018. pp. 36:1-36:5. DOI: 10.1145/3286606.3286813

[19] Yaakoubi AE, Amhaimar L, Attari K, Harrak MH, Halaoui ME, Asselman A. Non-linear and intelligent maximum power point tracking strategies for small size wind turbines: Performance analysis and comparison. Energy Reports. 2019;**5**:545-554. DOI: 10.1016/j.egyr.2019.03.001

[20] El Yaakoubi A, Attari K, Asselman A, Djebli A. Novel power capture optimization based sensorless maximum power point tracking strategy and internal model controller for wind turbines systems driven SCIG. Frontiers in Energy. 2017. DOI: 10.1007/s11708-017-0462-x (Article in press)

[21] Yaakoubi AE, Attari K, Amhaimar L, Asselman A. Adaptive state feedback pitch angle control of wind turbines for speed regulation and blades loadings alleviation. International Review of Automatic Control IREACO. 2018;**11**:174-187. DOI: 10.15866/ireaco.v11i4.14503

[22] El Yaakoubi A, Attari K, Asselman A, Djebli A. Simulation and investigation of the behavior of a large-scale direct driven wind turbine connected to the grid. UPB Scientific Bulletin, Series C: Electrical Engineering. 2017;**79**:241-252. DOI: 10.1016/j.egyr.2016.10.004

[23] Yaakoubi AE, Asselman A, Djebli A, Aroudam EH. A MPPT strategy based on fuzzy control for a wind energy conversion system. Procedia Technology. 2016;**22**:697-704. DOI: 10.1016/j.protcy.2016.01.145

Chapter 11

Robust Nonlinear Control Strategy for Small Wind Turbines: A Case Study

Ridha Cheikh and Hocine Belmili

Abstract

This chapter presents a case study of robust nonlinear control strategy using nonlinear feedback control technique based on Lyapunov theory and associated with robust control laws. The proposed approach aims to enhance robustness of the wind turbine control scheme. In fact, we selected as a case study, the most used electrical generator in small-scale wind applications, the Permanent Magnet Synchronous Generator (PMSM). Indeed, the control strategy presented in this chapter allows an efficient operation of the wind turbine in the standalone operating mode, offers a nonlinear handling of the WECS(s) and guarantees maximum wind power harvesting and robustness against critical working conditions. Talking about stability, in several wind generator control schemes; a such classical PI controllers-based scheme can easily be disturbed by any uncertainty of the system parameters, thus, in this chapter, we focused on how to overcome this issue by proposing a robust control strategy based on nonlinear controller derived from the Lyapunov Theory. The chapter presents numerical simulations within Matlab/SIMULINK environment. These results proved the effectiveness and the benefits of the proposed approach.

Keywords: WECS, PMSG, eigenvalues, nonlinear feedback control, Lyapunov theory, robust control

1. Introduction

During the last decade, the world has known a significant increase in electricity production from renewable energy sources, due in part to the liberalization of the electricity market, which has attracted the greed of new producers who are oriented towards cheaper productions (cogeneration, biomass, wind, etc.) and not in overly heavy investments (thermal power plants), and, secondly, because the emerging ecological awareness on climate change due to the emission of greenhouse gases has resulted in a political will that results from international protocols, such as Kyoto 1997 and Paris 2015, to encourage the use of clean and renewable energies for electricity generation [1].

Currently, wind energy is an important part of renewable energy production. Based on statistics from the Global Wind Energy Council (GWEC), more than 50 GW of clean, emission-free wind power was installed in 2018, bringing total installations to 591 GW globally; those statistics forecast that the global installed capacity can reach more than 817 GW by 2021 [2]. However, as the wind energy is very different from conventional sources, it is an intermittent energy source that is not continuously available for conversion into electricity and outside direct control because the used primary energy cannot be stored, it may be predictable but cannot be dispatched to meet the demand of an electric power system. Thus, the rapid growth of wind energy penetration into power systems causes many problems regarding the power flow control flexibility, which has forced many countries to revise their grid codes to ensure stable and reliable network operation. Power systems can reach very high wind penetration levels, as is the case of Denmark (42.4%), Germany (14.8%) and Spain (14.1%) [3, 4].

Nowadays the major challenge of the wind energy industry is to respect the deferent electrical grid requirements of each country; this means that power systems doest accept except a pure and balanced electrical energy comes from wind farms and in case of grid disturbances; those wind farms have to remain connected and support the grid to avoid black outs [5, 6]. As a result, the successful installation of wind turbines subject to requirements imposed by any country's energy market will not be possible without the contribution of advanced and robust control strategies [7].

Currently, wind turbine technology is almost based on two topologies: fixed speed-based topology (FST) which uses Induction Machines (IMs) and variable speed-based topology (VST) which generally uses Double Fed Induction Machines (DFIMs) or Permanent Magnet Synchronous Machines (PMSMs) [8].

However, the FST topology is recognized by many advantages such as costless and no complexity even more, nevertheless it has many disadvantages such as lower efficiency; power fluctuations in the grid due to wind gusts, short life cycle due to significant efforts undergone by its structure [9]. Thus, even the wind system complexity will increase, but VST, in which power fluctuations could be reduced and hence wind gusts can be stored as kinetic energy in large rotating masses, has been proposed to mitigate all drawbacks of FST.

In recent years, choosing the type of the electrical generator of WTs is a serious subject of several research works. This component represents the heart of a WT; it is used to produce electric power with lower cost and under variable nature of the wind speed. In the wind industry, two types of generators dominate the sector: (a) the Permanent Magnet Synchronous Generator (PMSG) with very low inertia, high volumetric torque, good efficiency and better controllability, which, furthermore, with the possibility to direct drive train, has become a serious competitor of (b) the Double Fed Induction Machines (DFIG). This latest (DFIG) is the most used in wind industry sector due to the ability to control powers flow with significantly improved yield. The DFIG can operate in different speed modes (sub-synchronous, synchronous and super-synchronous). Due to the progress of modern technologies, the use of the DFIG in large power scales seems to have become a more efficient solution for electrical energy generation whether for onshore or offshore installations [10–12].

Therefore, developing robust nonlinear control algorithms for electromechanical systems actuated by different types of electric machinery, emphasizing system stability and robustness, is a very essential issue [13]. Under this scope, this chapter proposes a robust nonlinear control strategy based on Lyapunov Theory of a standalone Permanent Magnet Synchronous Generator (PMSM). Therefore, the chapter is focused on the design of nonlinear control strategy using the nonlinear feedback control techniques and Lyapunov theory to guarantee robustness of the wind energy conversion systems.

Finally, based on simulation results, a general conclusion is presented in this chapter showing the performance of the proposed control strategy used for the studied wind turbine.

DOI: http://dx.doi.org/10.5772/intechopen.90750

2. Lyapunov theory and robust control association

2.1 Lyapunov theory fundamentals

The theory introduced in the late 19th century by the Russian mathematician Alexandr Mikhailovich Lyapunov is generally the most useful approach for analyzing the stability and designing control of linear or nonlinear systems. In fact, Lyapunov's work for stability problems includes two methods: the so-called linearization method and the direct method. The linearization method draws conclusions about a nonlinear system's local stability around an equilibrium point from the stability properties of its linear approximation [14]. However, the direct method (or second method) is not restricted to local motion, and determines the stability properties of a nonlinear system by constructing a scalar "energy-like" function for the system and examining the function's time variation [15]. In other words, the direct method is a generalization of the idea that if there is some "measure of energy" in a system, then we can study the rate of change of the system's energy to ascertain stability [16]. In this chapter, we will focus on the direct method for the control scheme design.

2.1.1 Lyapunov's direct method concept

Lyapunov's direct method for stability is now referred to as the Lyapunov stability criterion and makes use of a scalar "energy-like" function $V(x)$, which has an analogy to the potential function of classical dynamics.

It can be introduced for a system $\dot{x} = f(x)$ having a point of equilibrium at $x = 0$. Consider a function $V(x): R^n \to R$ such that

- $V(x) = 0$ if and only if $x = 0$. $V(x) > 0$ if and only if $x \neq 0$.
- $\dot{V}(x) = \frac{d}{dt}V(x) = \sum_{i=1}^{n} \frac{\partial V}{\partial x_i} f_i(x) = \nabla V.f(x) \leq 0$ for all values of $x \neq 0$.

For asymptotic stability, the condition $\dot{V}(x) < 0$ for $x \neq 0$ is required.

Then $V(x)$ is called a Lyapunov function and the system is stable in the sense of Lyapunov.

2.1.2 Stability of a linear system

Consider a linear system described by its state space model as follows:

$\dot{X} = AX$ (A is a finite matrix), we can say that A is asymptotically (or exponentially) stable if all its eigenvalues' real parts are negative; in other words, all the system states will converge to the vicinity of their equilibrium points [16].

The above stability condition can be verified equivalently through Lyapunov concept, using the quadratic Lyapunov function as follows:

$V = X^t PX$ with P is a symmetric positive definite matrix,

- If $X \neq 0 \to X^t PX > 0$
- then $\dot{V} = \dot{X}^t PX + X^t P\dot{X}$. Replacing the value of $\dot{X}$ we have the following:

$$\dot{V} = X^t(A^t P + PA)X = -X^t QX, \; so, \; A^t P + PA = -Q$$

The stability condition is that $\dot{V} < 0$ so the matrix Q should be positive definite for some definite matrix $P = P^t$ [17].

2.2 Modeling of parametric uncertainty for AC machines

2.2.1 Resistance variation

Heating of the machine windings due to temperature increase causes an increase in stator and rotor resistances. This variation can be generally presented in the following relation:

$$R = R_0(1 + \alpha T)$$

where α is the temperature coefficient value ($\approx 4.1 \times 10^{-3}$) and T is the temperature in degree Celsius.

2.2.2 Inductance variation

The variation of the inductances is caused by the saturation of the ferromagnetic materials of the machine. The dynamics of this variation and that of currents cannot be considered slow [18].

2.2.3 Skin effect

This concerns the induction machine; in fact, the rotor bars' resistance increases with the value of the frequency of rotor currents. An increase in the frequency causes a backflow of the currents to the periphery of the bars. The variation of the resistance of the bars is calculated numerically using a finite element method [18].

In steady-state operation, that is to say in the absence of parametric disturbances, the machine parameters take nominal values.

In general, the machine parameters are subject to unknown variations resulting from the different situations in which it evolves. The variation of the resistances is directly related to the fluctuations of the temperature of the machine; whereas the variations of the inductances are related to the phenomenon of magnetic saturation. This type of disturbance is called structured uncertainty.

From above, let us express the electrical parametric variations of the machine, such as:

$$\begin{cases} R_{s,r} \rightarrow R^0_{s,r} + \Delta R_{s,r}, L_{s,r} \rightarrow L^0_{s,r} + \Delta L_{s,r} \\ \sigma \rightarrow \sigma^0 + \Delta\sigma, T_r \rightarrow T^0_r + \Delta T_r \end{cases}$$

2.2.4 Global uncertainty model

For an AC machine represented by the nonlinear system below:

$$\begin{cases} \dot{x} = f(x) + g(x)u \\ y = h(x) \end{cases}$$

To globally model uncertainty of the previous system, we can rewrite it as follows:

$$\begin{cases} \dot{x} = f(x) + \Delta f(x) + (g(x) + \Delta g(x))u \\ y = h(x) \end{cases}$$

where the exact part of the plant is represented by the functions (f, g, h), while (Δf, Δg) represent the uncertain part of the system [19].

A robust control of the uncertainty is not guaranteed by the linear control law. Thus, one should design a robust law based on the Lyapunov theory to achieve an input–output stability and robustness against uncertainty [19].

3. Case study: robust control of a standalone WECS-based PMSG

3.1 System modeling

The schema of the real studied system is presented in **Figure 1(a)**. However, for simplification purposes, and due to the power electronics and local grid dynamics being much faster than other system dynamics, it will not be considered and the simplified system is presented in **Figure 1(b)** [14].

From **Figure 1(b)**, the control goal is to change the chopper equivalent resistance R_L (load) at the generator terminals in order to adjust the generator speed to its optimum value, and then the maximum power capture will be achieved.

The global system model can be expressed part per part in the following equations:

$$\left\{ \begin{array}{l} v(t) = \overline{v} + \Delta v(t), \\ P_t = \Gamma_t \omega_r \\ \Gamma_t = \dfrac{1}{2}\pi\rho R^3 v^2 C_\Gamma(\lambda) \\ C_\Gamma(\lambda) = q_2\lambda^2 + q_1\lambda + q_0 \\ \lambda = \dfrac{\omega_r R}{v} \\ C_P = \lambda \times C_\Gamma \end{array} \right| Aero-mechanicalpart \quad (1)$$

$$\left\{ \left\{ \begin{array}{l} \dfrac{d}{dt} i_d = -\dfrac{R_s + R_L}{L_d + L_L} i_d + \dfrac{p(L_q - L_L)}{L_d + L_L} i_q \omega_h \\ \dfrac{d}{dt} i_q = -\dfrac{R_s + R_L}{L_q + L_L} i_q - \dfrac{p(L_d + L_L)}{L_q + L_L} i_d \omega_h + \dfrac{p\Phi_m}{L_q + L_L}\omega_h \end{array} \right| electricalpart \right. \quad (2)$$

$$\left\{ \left\{ \begin{array}{l} J_h \dfrac{d\omega_h}{dt} = \dfrac{\eta}{i}\Gamma_t - \Gamma_g \\ \omega_h = \omega_r \times i \\ \Gamma_g = p\Phi_m i_q \end{array} \right| Electro-mecanicalpart \right. \quad (3)$$

where $v(t)$ is the wind speed (average and turbulent components); (Γ_t, Γ_g) are respectively the turbine and generator torques; P_t is the mechanical power; (ω_r, ω_h) are turbine and generator speeds respectively; ρ is the air density; R is the swept surface radius; $C_\Gamma(\lambda)$ is the torque coefficient; (q_0, q_1, q_2) are given in the appendix; C_P is the power coefficient; λ is the tip speed ratio (TSR); η is the efficiency; i is the

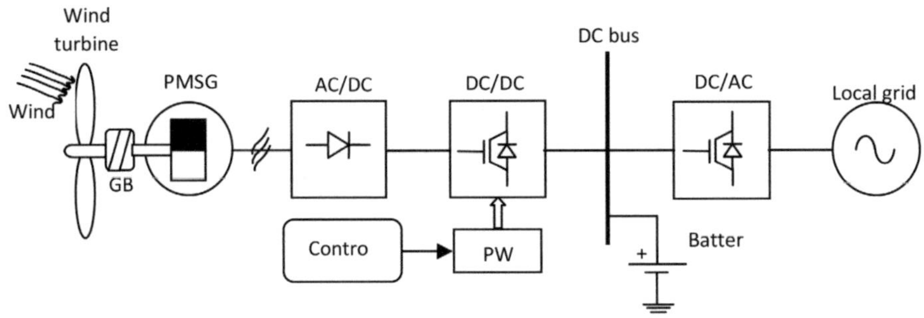

(a)

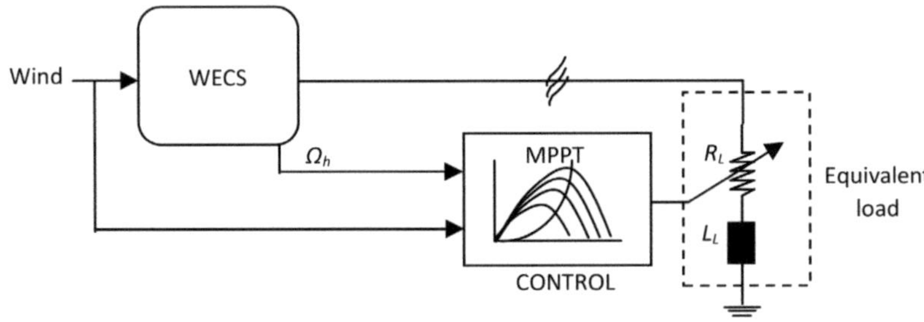

(b)

Figure 1.
Schema of the studied WECS-based PMSG: (a) real system; (b) simplified system.

gearbox ration; i_d, i_q, L_d, L_q are respectively the d/q axis currents and inductances; Rs is the stator resistance; p is the pair poles; R_L and L_L are respectively the resistance and inductance of the equivalent load.

From Eq. (1) to Eq. (3), choosing radius $x = [x_1, x_2, x_3]^T = [i_d, i_q, \omega_h]^T$ as state vector, $u = [u_1, u_2] = [R_L, v]$ as control signal and $y = \omega_h$ as a desired output, the nonlinear state space model of the system can be defined as follows:

$$\left\{ \dot{x} = \begin{bmatrix} \frac{1}{L_d + L_L}\left(-R_s x_1 + p\left(L_q - L_L\right)x_2 x_3\right) \\ \frac{1}{L_q + L_L}\begin{pmatrix} -R_s x_2 - p(L_d + L_L)x_1 x_3 \\ +p\Phi_m x_3 \end{pmatrix} \\ \frac{1}{J_h}\left(d_3 x_3^2 - p\Phi_m x_2\right) \end{bmatrix} + \begin{bmatrix} -\frac{1}{L_d + L_L} x_1 u_1 \\ -\frac{1}{L_q + L_L} x_2 u_2 \\ \frac{1}{J_h}\left(d_1 u_2^2 + d_2 u_2 x_3\right) \end{bmatrix} \right. \tag{4}$$

where $d_1 = \frac{\eta\rho\pi R^3}{2i} q_0, d_2 = \frac{\eta\rho\pi R^4}{2i^2} q_1, d_3 = \frac{\eta\rho\pi R^5}{2i^3} q_2.$

3.2 Lyapunov linearization method

In this subsection, we will apply the Lyapunov Linearization (first method) to the studied system (Eq. (4)) in order to deal with the system's small-signal stability under wind speed variation.

Suppose that $(\overline{x}, \overline{u})$ is an equilibrium point and input [20].

Using Taylor's series expansion of the function f we have:

$$\begin{cases} \delta x = x - \overline{x} = [\delta x_1 \delta x_2 \delta x_3]^T \\ \delta u = u - \overline{u} = [\delta u_1 \delta u_2]^T \end{cases} \tag{5}$$

The derivative variable within a unique function can be written as:

$$\dot{x} = F(x(t), u(t)) \tag{6}$$

$$\dot{\delta}_x(t) \approx F(\overline{x}, \overline{u}) + \left.\frac{\partial F}{\partial x}\right|_{\substack{x = \overline{x} \\ u = \overline{u}}} \delta_x(t) + \left.\frac{\partial F}{\partial u}\right|_{\substack{x = \overline{x} \\ u = \overline{u}}} \delta_u(t) + \underbrace{\text{h.o.t}}_{0} \tag{7}$$

where $F(\overline{x}, \overline{u}) = 0$.
Then:

$$\dot{\delta}_x(t) \approx \left.\frac{\partial F}{\partial x}\right|_{\substack{x = \overline{x} \\ u = \overline{u}}} \delta_x(t) + \left.\frac{\partial F}{\partial u}\right|_{\substack{x = \overline{x} \\ u = \overline{u}}} \delta_u(t) \tag{8}$$

The matrices, A, B of the LTI system are constant matrices:

$$A \approx \left.\frac{\partial F}{\partial x}\right|_{\substack{x = \overline{x} \\ u = \overline{u}}} \in R^n \times R^n, B \approx \left.\frac{\partial F}{\partial u}\right|_{\substack{x = \overline{x} \\ u = \overline{u}}} \in R^n \times R^m$$

The linearized version around an equilibrium point characterized by the quintuple $[\overline{x}, \overline{u}] = [\overline{i}_{sd} \overline{i}_{sq} \overline{\omega}_h \overline{R}_L \overline{v}]$ can be written as follows:

$$\dot{\delta}_x = \begin{bmatrix} a_1 + a_2 \overline{R}_L & a_3 \overline{x}_3 & a_3 \overline{x}_2 \\ b_1 \overline{x}_3 & b_2 + b_3 \overline{R}_L & b_1 \overline{x}_1 + b_4 \\ 0 & c_4 & c_2 \overline{v} + c_3 \overline{x}_3 \end{bmatrix} \delta_x + \begin{bmatrix} a_2 \overline{x}_1 & 0 \\ b_3 \overline{x}_2 & 0 \\ 0 & c_1 \overline{v} + c_2 \overline{x}_3 \end{bmatrix} \delta_u \tag{9}$$

where,

$$\begin{cases} a_1 = -\dfrac{R_s}{L_d + L_L}; a_2 = -\dfrac{1}{L_d + L_L}; a_3 = \dfrac{p(L_q - L_L)}{L_d + L_L}. \\ b_1 = -\dfrac{p(L_d + L_L)}{L_q + L_L}; b_2 = -\dfrac{R_s}{L_q + L_L}; b_3 = -\dfrac{1}{L_q + L_L}; \\ b_4 = \dfrac{p\Phi_m}{L_q + L_L}; c_1 = \dfrac{2d_1}{J_h}; c_2 = \dfrac{d_2}{J_h}; c_3 = \dfrac{2d_3}{J_h}; c_4 = -\dfrac{p\Phi_m}{J_h}. \end{cases}$$

3.2.1 Small-signal stability

In this section, the small-signal stability of the linearized system will be checked regarding wind speed variation; therefore, we will have a point of view about the impact of the wind speed variation on the system stability. However, in open loop, without control, the system is excited by a wide range of wind speed variation starting from the cut-in to cut-out wind speed. The eigenvalues of the matrix A (Jacobian) are calculated and presented in **Table 1**.

$\bar{v}$ (m/s)	4	5	6	7	10	12
$\lambda_{1,2}$	−8.56±15i	−7.09±18i	−6.29±20i	−5.74±23i	−5.07±32i	−4.95±37i
λ_3	−279	−342	−406	−470	−666	−798

Table 1.
Linearized system eigenvalues under wind speed variation.

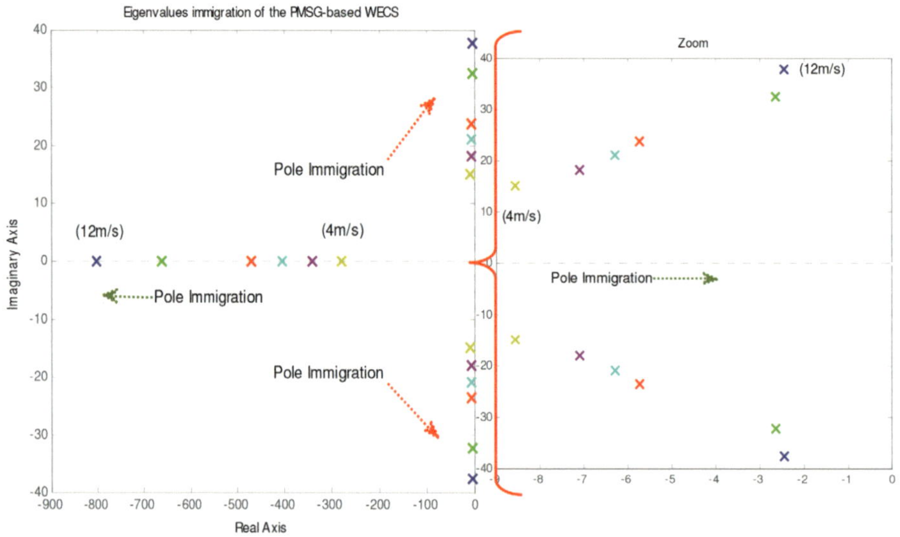

Figure 2.
Eigenvalues (poles) immigrating under wind speed variation.

Figure 2 shows clearly that for all values of wind speed, all the eigenvalues still have negative real parts and take places in the left-half of the complex plane, which means the system is almost stable under small perturbation. However, with wind speed increase, the complex-conjugated eigenvalues start immigrating almost close to the imaginary axis and then exhibit bad damping characteristics (**Figure 3**). Hence, damping should be improved in closed loop-based linear controller.

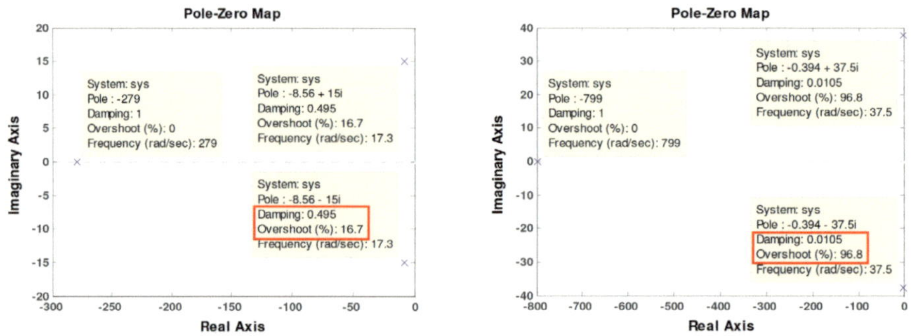

Figure 3.
System characteristics regarding damping/overshoot for two wind speed cases (4 m/s left, 12 m/s write).

3.3 System nonlinear feedback linearization

If we assume the wind speed dynamics to be very slow compared to the other system dynamics, the nonlinear system of Eq. (4) can be reformulated as follows:

$$\begin{cases} \dot{x} = \underbrace{\begin{bmatrix} -\dfrac{R_s}{L_d+L_L}x_1 + \dfrac{p(L_q-L_L)}{L_d+L_L}x_2x_3 \\ -\dfrac{R_s}{L_q+L_L}x_2 - \dfrac{p(L_d+L_L)}{L_q+L_L}x_1x_3 + \dfrac{p\Phi_m}{L_q+L_L}x_3 \\ \dfrac{1}{J_h}\left(d_1v^2+d_2vx_3+d_3x_3^2-p\Phi_m x_2\right) \end{bmatrix}}_{f(x)} + \underbrace{\begin{bmatrix} -\dfrac{1}{L_d+L_L}x_1 \\ -\dfrac{1}{L_q+L_L}x_2, 0 \end{bmatrix}}_{g(x)} u \\ y = h(x) = x_3 \end{cases} \tag{10}$$

where $d_1 = \frac{\eta\rho\pi R^3}{2i}q_0, d_2 = \frac{\eta\rho\pi R^4}{2i^2}q_1, d_3 = \frac{\eta\rho\pi R^5}{2i^3}q_2$.

The system of Eq. (10) has smooth nonlinearities; therefore, the Feedback Linearization Technique can be applied to control the system.

From Eq. (10), taking derivatives of output y with respect to time up to r times will give:

$$y^{(r)} = L_f^r h(x) + \underbrace{L_g L_f^{r-1} h(x) u}_{\neq 0} \tag{11}$$

r is the relative degree ($r < n$).

$L_f^r h(x)$ is the Lie derivative of $h(x)$ along the direction of the vector field $f(x)$ up to r times, $L_g L_f^{r-1} h(x)$ is the Lie derivative of h(x) along the direction of the vector field $g(x)$.

$$\text{If we have} \begin{cases} \alpha(x) = L_f^r h(x) \\ \beta(x) = L_g L_f^{r-1} h(x) \end{cases} \tag{12}$$

Eq. (11) becomes:

$$y^{(r)} = \alpha(x) + \beta(x)u \tag{13}$$

Then the feedback linearization control can be chosen as follows:

$$u^*(x) = \frac{1}{\beta(x)}(-\alpha(x) + v) \tag{14}$$

$$y^{(r)} = v \tag{15}$$

The relative degree of Eq. (20) is r = 2, it is a linear input–output double-integrator, using linear methods to guarantee the system control (tracking problem).

Note that, the field vectors $\alpha(x)$ and $\beta(x)$ should be completely known to apply the feedback linearization technique [21].

The computation of the field vectors $\alpha(x)$ and $\beta(x)$ gives:

$$\begin{cases} \alpha(x) = -\frac{p\Phi_m}{J_h}\left[b_2x_2 + b_1x_1x_3 + b_4x_3\right] + \frac{1}{J_h}\left[\frac{d_2}{i}v + 2d_3x_3\right]\dot{x}_3 \\ \beta(x) = \frac{b_4}{J_h}x_2 \end{cases} \tag{16}$$

To harvest the maximum wind power, the generator should be running at optimal speed for each wind speed, thus the generator speed reference can be expressed as follows:

$$y_r(t) = \omega_h^* = \frac{\lambda^* iv(t)}{R} \tag{17}$$

where λ^* is the optimal tip speed ratio.
Let us take an output tracking error as follows:

$$e(t) = y_r(t) - y(t) \tag{18}$$

and introduce the following error vector:

$$\xi(t) = \left[e(t) \ \ \dot{e}(t)\right]^T \tag{19}$$

Hence, the control objective can be achieved using the ideal control law:

$$u^*(x) = \frac{1}{\beta(x)}\left(-\alpha(x) + y_r^{(2)}(t) - k^T\xi\right) \tag{20}$$

The polynomial $s^2 + k_1s + k_2$ would have all its roots in the left-half of the complex plane if we choose an appropriate gain matrix $k = \left[k_2 \quad k_1\right]^T$. This means that the tracking error asymptotically converges to zero ($\ddot{e}(t) + k_1\dot{e}(t) + k_2e(t) = 0$). Thus, the system poles are placed following the specifications concerning the overshoots and the settling times.

3.4 Robust control law design

As we mentioned previously, AC machines are subject to a large model uncertainty due to parameter variations, noises, measurement errors etc. In this chapter, we will take into account just electrical and mechanical parametric uncertainties. In the studied case, it is clear that a successful control strategy of the WECS-based PMSG hinges on the good computing of control u, which is strongly dependent on the machine parameters (Eq. (20)). Therefore, we propose a novel robust control law to handle the uncertainty issue and offer an accurate description of the system model [19].

From Section 2.2.1, Eq. (10) can be rewritten as follows:

$$\begin{cases} \dot{x} = f(x) + \Delta f(x) + (g(x) + \Delta g(x))u \\ y = h(x) \end{cases} \tag{21}$$

However, the exact parts of the model are $f(x)$, $g(x)$ and $h(x)$, while Δf and Δg present the uncertainty.

Consequently, we have:

$$\begin{cases} \ddot{y} = v + \Delta v \\ \Delta v = \Delta\alpha(x) + \Delta\beta(x)u \end{cases} \tag{22}$$

Δv is the effect of uncertainty, it is an unknown term, which makes the linear feedback control law of Eq. (15) incapable of providing a robust tracking performance of the system. Thus, the following robust control law is proposed to handle this issue.

From Eq. (22), we rewrite the error dynamics as follows:

$$\dot{\xi} = \underbrace{\begin{bmatrix} 0 & 1 \\ -k_2 & -k_1 \end{bmatrix}}_{A_c}\xi + \underbrace{\begin{bmatrix} 0 \\ 1 \end{bmatrix}}_{B}(v + \Delta v) \tag{23}$$

We chose the Lyapunov function as follows:

$$V = \frac{1}{2}\xi^T P\xi \tag{24}$$

The derivation of V gives

$$\dot{V} = \frac{1}{2}\left[\dot{\xi}P\xi + \xi^T P\dot{\xi}\right] \tag{25}$$

P is a positive semi-definite matrix in which

$$A_c{}^T P + PA_c = -Q \tag{26}$$

Q is an identity matrix.

After computation, $\dot{V}$ becomes

$$\dot{V} = -\frac{1}{2}\xi^T Q\xi + \xi^T PB(v + \Delta v) \tag{27}$$

V satisfies the condition $\dot{V} < 0$ along the solution trajectory of the system, which gives the following:

$$\xi^T PB(v + \Delta v) < 0 \tag{28}$$

v is chosen as

$$v = -F\,\mathrm{sgn}\left(\xi^T PB\right) \tag{29}$$

with $F > |\Delta v|$.

The substitution of Eq. (29) in Eq. (28) gives

$$\xi^T PB\left(-F\,\mathrm{sgn}\left(\xi^T PB\right) + \Delta v\right) < 0 \tag{30}$$

Eq. (30) is always negative for any value of $\xi^T PB$ provided $F > |\Delta v|$. The value of the gain F is chosen accordingly by trial and error.

We have

$$\mathrm{sgn}\left(\xi^T PB\right) = \begin{cases} 1 & if \quad \xi^T PB > 0 \\ 0 & if \quad \xi^T PB = 0 \\ -1 & if \quad \xi^T PB < 0 \end{cases} \tag{31}$$

After computation, the P matrix element can be set as follows:

$$P = \begin{bmatrix} \frac{K_1}{2K_2} + \frac{1+K_2}{2K_1} & \frac{1}{2K_2} \\ \frac{1}{2K_2} & \frac{1+K_2}{2K_1K_2} \end{bmatrix} \tag{32}$$

Finally, the linear control signal is expressed as

$$\upsilon = y_r^{(2)}(t) - k^T\xi - F\,\mathrm{sgn}\left(\left[\frac{1}{2K_2}\ \frac{1+K_2}{2K_1K_2}\right]\xi\right) \tag{33}$$

4. Simulation results and discussion

Based on the control scheme shown in **Figure 4**, we have simulated within Matlab/Simulink® the studied system shown in **Figure 1**. The system data are given in the appendix. Through simulation, we have considered that all system variable states (x_1, x_2, x_3) are available for feedback. The simulation aims to check two control performances, a wind maximum power tracking performance under a realistic wind speed, and robustness performance against both rapid wind speed variation (sharp) and parametric uncertainty of the generator inertia.

4.1 Tracking performance check

For tracking performance under wind speed fluctuations (**Figure 5(a)–(d)**), the power coefficient C_p holds easily its maximum value (C_{pmax}) and the same for tip speed ratio, TSR (λ^*). Consequently, the PMSG runs at optimal speeds for each wind speed value (see **Figure 5(d)**) and the maximum power extraction is indeed guaranteed (**Figure 6(b)**).

However, from **Figure 6(a)**, it should be mentioned that the control signal, which represents the equivalent chopper resistance (R_L), has the chattering phenomenon due to discontinuous control effect (see Eq. (33)). From **Figure 6(c)** and **(d)**, all system operating points are in the neighborhood of the optimal operating points.

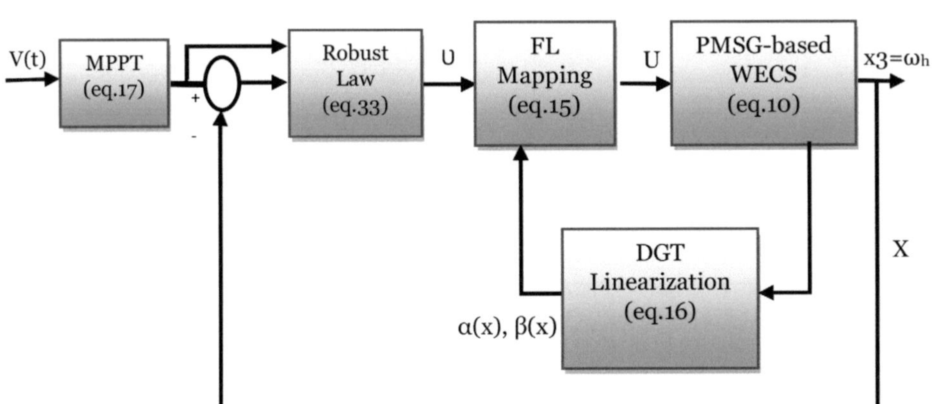

Figure 4.
Robust control simulation scheme.

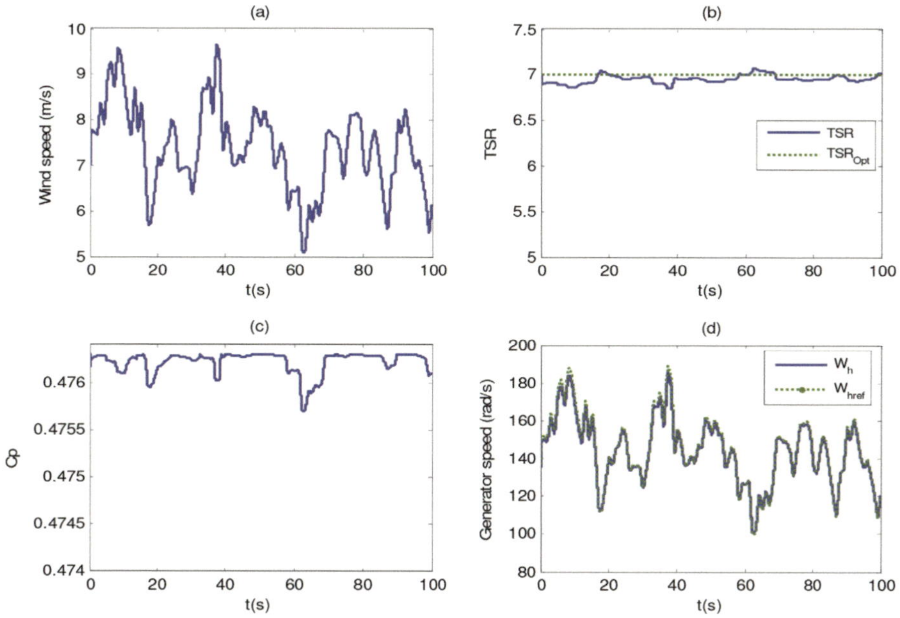

Figure 5.
Check of tracking performance under a realistic wind speed. (a) Wind speed; (b) TSR; (c) power coefficient; (d) generator speed.

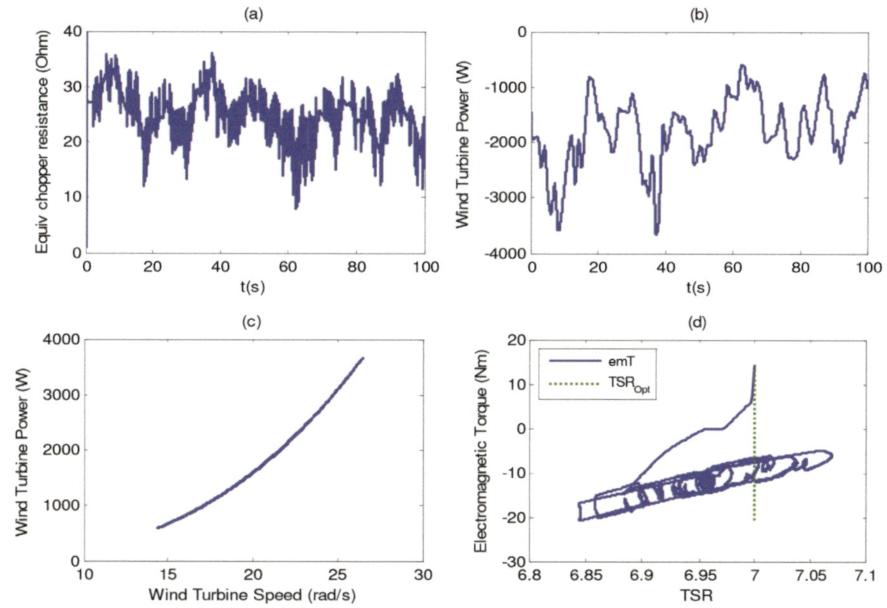

Figure 6.
(a) Control signal; (b) turbine power; (c) optimal points neighborhood; (d) electromagnetic torque versus TSR.

Then, the next simulation is to check the performance of the proposed controller facing critical wind speed conditions. Hence, we have subjected the system to a sharp rise and drop variation of the wind speed (**Figure 7(a)**). The simulation results show that even under abrupt changes of the wind speed, the system kept its stability and continued working after some short transitory disturbances and small static errors, which can be observed through **Figure 7(a)**-**(c)** and **Figure 8(a)**-**(c)**.

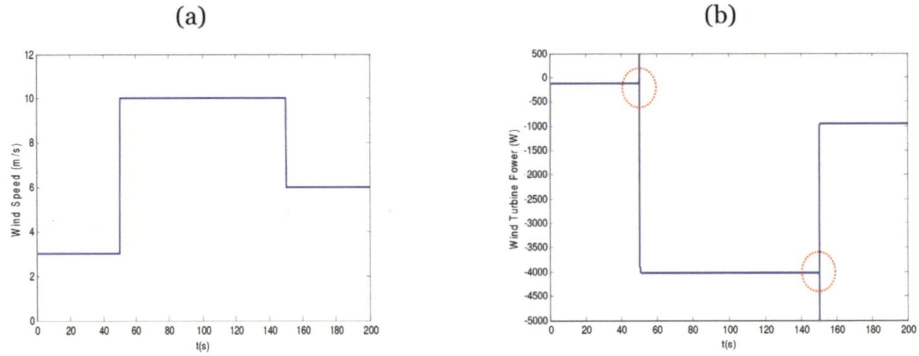

Figure 7.
Robustness check under sharp wind speed variation. (a) Wind speed profile; (b) wind turbine electrical power.

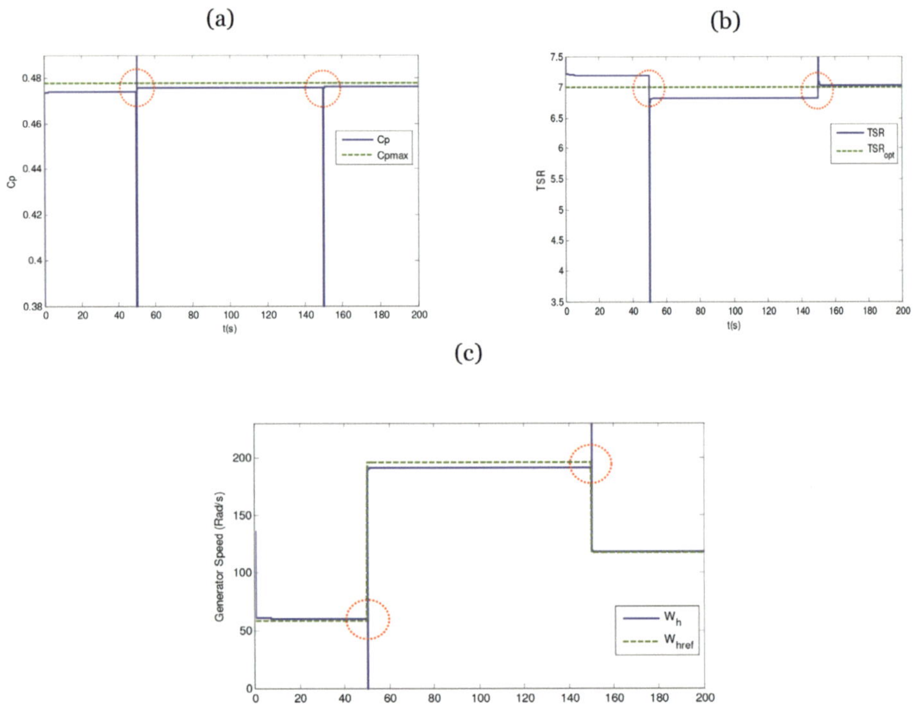

Figure 8.
(a) Power coefficient; (b) tip speed ratio; (c) generator speed.

4.2 Parametric uncertainty robustness check

Following Eq. (14) and Eq. (16), it is obvious that the control signal is calculated based on field vectors $\alpha(x)$ and $\beta(x)$ whose values are strongly dependent on the system parameters, especially the high speed shaft inertia J_h. Thus, in order to check the proposed controller robustness against this parametric uncertainty, we have simulated the system with sharp variation of high speed shaft inertia as shown in **Figure 9(a)**. From the obtained simulation results (**Figures 10** and **11**), except small static errors and very small disturbances, this parametric uncertainty would not

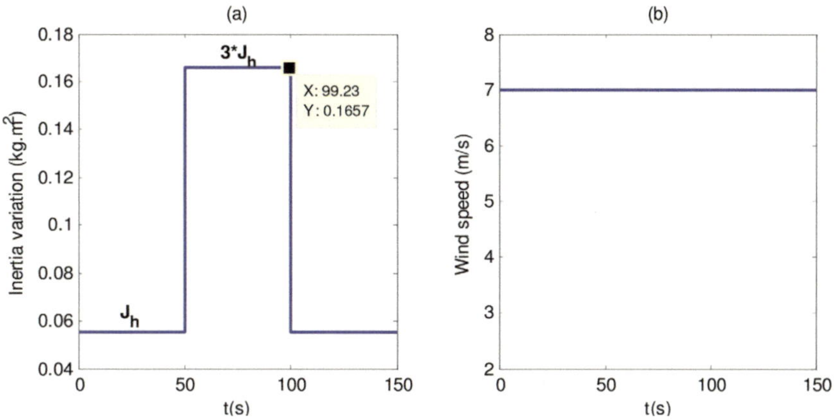

Figure 9.
(a) High speed shaft inertia; (b) average wind speed.

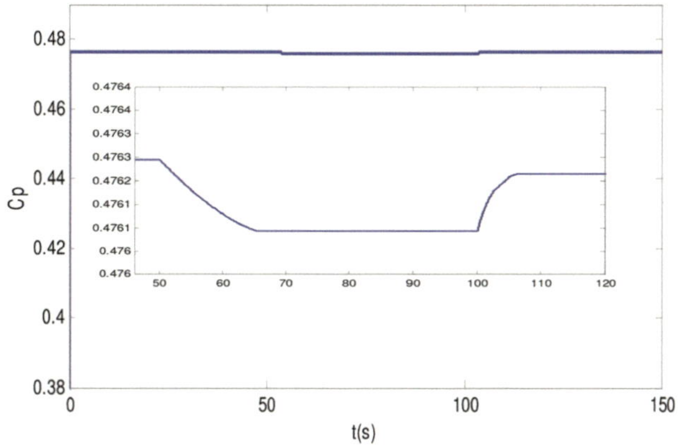

Figure 10.
Power coefficient C_p.

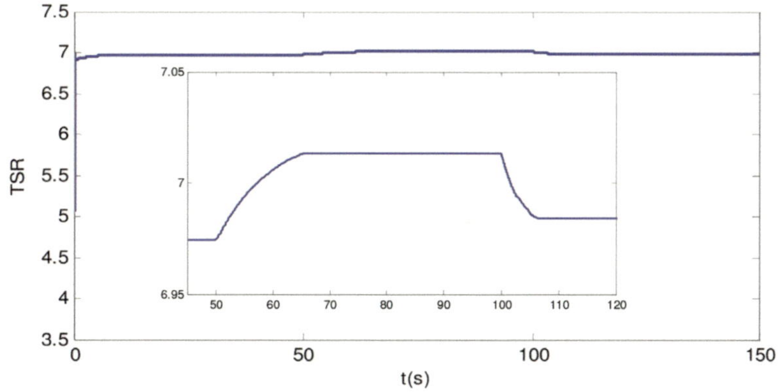

Figure 11.
Tip speed ratio.

have any notable effect on the stability of the control scheme. However, the field vectors $\alpha(x)$ and $\beta(x)$ dynamics, the control signal and the generator speed can be shown in and **Figure 12**.

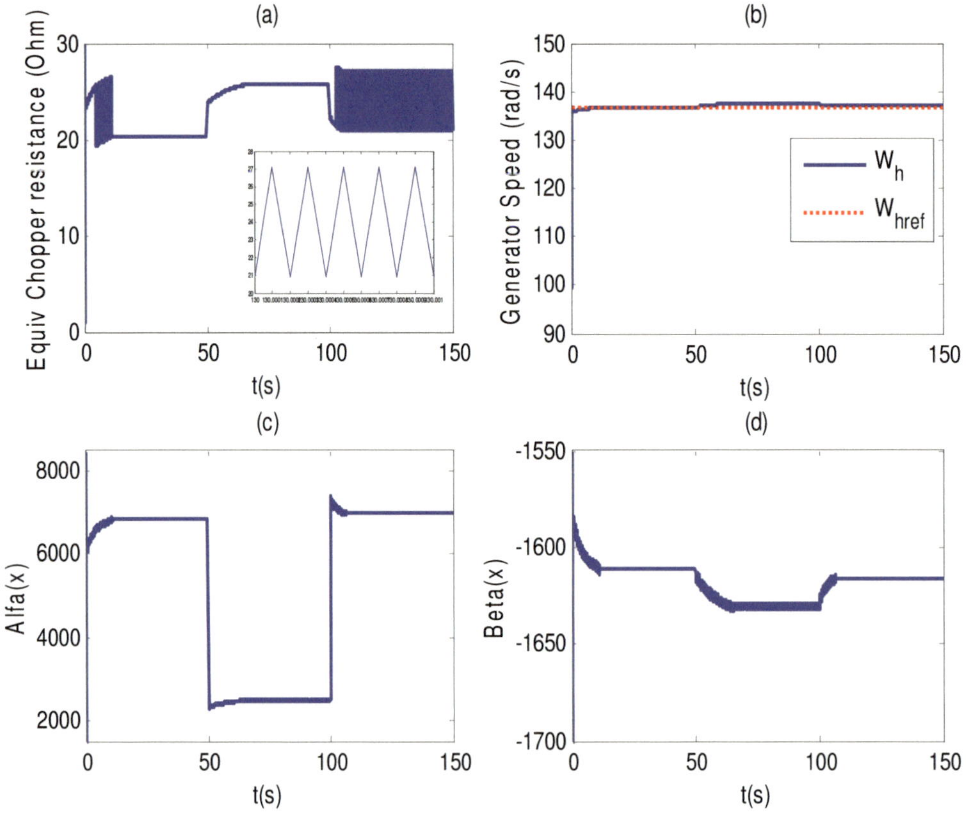

Figure 12.
(a) Control signal; (b) generator speed; (c) $\alpha(x)$; (d) $\beta(x)$.

5. Conclusion

In this chapter, we have discussed the robust control technique based on Lyapunov theory to ensure tracking performance and robustness of a nonlinear system. Mainly, the chapter presents this control strategy based on a case study of a standalone nonlinear WECS-based PMSG. In fact, we focused on tracking performance to ensure maximum wind power extraction and robustness performance against parametric uncertainty. Whereas, the control strategy uses the state feedback linearization technique associated with a control law derived from the Lyapunov theory. The obtained results through simulation have proved the efficiency of the proposed control technique.

Appendix

Turbine rotor:

$R = 2.5$ m $\rho = 1.25$ kg/m^3, $C_{Pmax} = 0.477$, $\lambda^* = 7$

Drive train:

$\eta = 1$, $i = 7$; $J_h = 0.0552$ kg m^2

Torque coefficient parameters:

$q_0 = 0.0061$, $q_1 = -0.0013$, $q_2 = -9.7477 \times 10^{-4}$

DOI: http://dx.doi.org/10.5772/intechopen.90750

PMSG:

$p = 3$, $R_s = 3.3\ \Omega$
$L_d = 0.0416$ H
$L_q = 0.0416$ H
$L_L = 0.08$ H
$\mathbf{\Phi}_m = 0.4382$ Wb
$V_s = 380$ V

Control parameters

Linear control: $k_1 = 100$. $k_2 = 4000$
Robust control: $F = 5000$.

Author details

Ridha Cheikh* and Hocine Belmili
Unité de Développement des Equipements Solaires, UDES, Centre de Développement des Energies Renouvelables, CDER, Tipaza, Algeria

*Address all correspondence to: cheikh_red@yahoo.fr

References

[1] Available from: https://unfccc.int/resource/bigpicture/#content-the-paris-agreement

[2] Global Wind Energy Council (GWEC). Global wind statistics. Report 2018. 2018. Available at: www.gwec.net [Accessed: April 2019]

[3] Flynn D et al. Advances in Energy Systems, Technical Impacts of High Penetration Levels of Wind Power on Power System Stability: The Large-scale Renewable Energy Integration Challenge. 1st ed. John Wiley & Sons Ltd; 2019. DOI: 10.1002/9781119508311.ch3

[4] Wind Europe. 2019. Available at: https://windeurope.org/about-wind/daily-wind/

[5] Geng H et al. LVRT capability of DFIG-based WECS under asymmetrical grid fault condition. IEEE Transactions on Industrial Electronics. 2013;**60**(6): 2495-2409

[6] Shuai Xiao, Hong lin Zhou. An LVRT control strategy based on flux linkage tracking for DFIG-based WECS. IEEE Transactions on Industrial Electronics. 2013;**60**(7):2820-2832

[7] Gupta A, Shandilya A. Challenges of integration of wind power on power system grid: A review. International Journal of Emerging Technology and Advanced Engineering. 2014;**4**(4): 880-884

[8] Poitier F. Study and control of asynchronous generator for the use of wind energy-stand-alone induction machine-grid-connected double fed induction machine [PhD thesis]. Polytechnic School of the University of Nantes; 2003

[9] El Aimani S. Modeling of different wind turbine technologies integrated into a medium voltage network [PhD thesis]. École Centrale of Lille; 2004

[10] Abad G et al. Doubly Fed Induction Machine Modeling and Control for Wind Energy Generation. 1st ed. IEEE-John Wiley & Sons, Inc.; 2011

[11] Pena R et al. Doubly fed induction generator using back-to-back PWM converters and its application to variable-speed wind-energy generation. IEE Proceedings—Electric Power Applications. 1996;**143**(3):231-241

[12] Cheikh R et al. Robust nonlinear control via feedback linearization and Lyapunov theory for permanent magnet synchronous generator-based wind energy conversion system. Frontier in Energy. 2018;**12**(46):1-12

[13] Dawson DM et al. Nonlinear Control of Electric Machinery. 1st ed. UK: Taylor & Francis Group; 1998. DOI: 10.1201/9780203745632

[14] Munteanu I et al. Optimal Control of Wind Energy Systems: Toward a Global Approach. UK: Springer-Verlag London Limited; 2008

[15] Slotine JJE, Li W. Applied Nonlinear Control. New Jersey: Prentice-Hall, Inc.; 1991

[16] Murray RM, Li Z, Sastry SS. A Mathematical Introduction to Robotic Manipulation. CRC Press; 1993

[17] Abdelmalek G. Commande Robuste d'un Dispositif FACTS par les Méthodes Métaheuristiques pour la Stabilité de Tension d'un Réseau Electrique [thesis]. Department of Electrical Engineering, Biskra University University; 2019

[18] Penman J, Sedding HG, Lloyf BA, Fink WT. Detection and location of interturns short circuits in the stator

windings of operating motors. IEEE Transactions on Energy Conversion. 1994;**9**(4)

[19] Ridha C et al. Robust control based on the Lyapunov theory of a grid-connected doubly fed induction. Frontiers in Energy. 2013;7(2):191-196. DOI: 10.1007/s11708-013-0245-y

[20] Tailor MR, Bhathawala PH. Linearization of nonlinear differential equation by Taylor's series expansion and use of Jacobian linearization process. International Journal of Theoretical and Applied Science. 2011; **4**(1):36-38

[21] Isidori A. Nonlinear Control Systems. 3rd ed. Berlin: Springer-Verlag; 1995